FEB 8 - 1996			

BIOGEOGRAPHICAL PROCESSES

I. G. Simmons

Department of Geography, University of Durham

London
GEORGE ALLEN & UNWIN
Boston   Sydney

George Allen & Unwin (Publishers) Ltd,
40 Museum Street, London WC1A 1LU, UK

George Allen & Unwin (Publishers) Ltd,
Park Lane, Hemel Hempstead, Herts HP2 4TE, UK

Allen & Unwin Inc.,
9 Winchester Terrace, Winchester, Mass 01890, USA

George Allen & Unwin Australia Pty Ltd,
8 Napier Street, North Sydney, NSW 2060, Australia

First published in 1982

British Library Cataloguing in Publication Data

Simmons, Ian
    Biogeographical processes. – (Processes in physical geography; 5)
1. Biogeography
I. Title   II. Series
574.9   QH84
ISBN 0-04-574016-X

Library of Congress Cataloging in Publication Data

Simmons, I. G. (Ian Gordon)
    Biogeographical processes.
(Processes in physical geography; 5)
Bibliography: p.
Includes index.
1. Biogeography. I. Title. II. Series.
QH84.S54   574.9   81-12666
ISBN 0-04-574016-X (pbk.) AACR2

Set in 10 on 11 point Times by Fotographics (Bedford) Ltd
and printed in Great Britain
by Fletcher and Son Ltd, Norwich

Preface

It is perhaps unusual for professors to write school books nowadays but I can assure my colleagues that it is a salutary and stimulating exercise. What has worried me is that, unlike the other books in this series, no practising school teacher could be found to write this book. One of my hopes is that by doing it myself I shall help to make sure that eventually such a situation is corrected.

My thanks are due to everybody who helped with the production, and especially to Mrs Macey and Mrs Southcott who did most of the typing. I am very grateful to Darrell Weyman, the general editor of this series, who provided a great deal of constructive comment on the early drafts. The other teachers to whom a draft was sent for review also provided a great deal of comment, much of it quite useful.

The text was written during a period of low teaching load made possible by an informal arrangement with my colleagues at Bristol University and my particular thanks go to Professor Peter Haggett, then Head of the Department, for making this possible.

I. G. SIMMONS
Bristol, December 1980

Acknowledgements

We gratefully acknowledge the following individuals and organisations who have supplied or given permission for the use of illustrative material:

Aerofilms Ltd; Heather Angel; P. Armstrong; Keith Barber; Barnaby's Picture Library; R. Barnes; Biophoto Associates; F. H. Bormann; the Botanical Society of the British Isles; Butterworths Ltd; J. L. Cloudsley-Thompson; Bruce Coleman Ltd; A. S. Collinson; J. E. Cousens; Eric Crichton; Croom Helm Ltd and J. L. Cloudsley-Thompson for permission to reproduce Fig. 15 from *Terrestrial environments;* J. Deedy; Duke University Press for permission to reproduce Fig. 6.10 from G. M. Woodwell and A. L. Rebuck, 'Effects of chronic gamma radiation on the structure and diversity of an oak–pine forest' (*Ecological Monographs* **37**, 53–69), © 1967 Ecological Society of America; Ecological Society of America for permission to reproduce Fig. 6.14 of *Dynamic ecology* by Collier *et al.* (1973), © 1973 Ecological Society of America; M. P. L. Fogden; Peter Fraenkel; M. Freeman; R. Good; Holt, Rinehart and Winston for permission to reproduce Fig. 14.7 from *Fundamentals of ecology*, 3rd edn, by Eugene P. Odum, © 1971 W. B. Saunders Company, © 1953, 1959 W. B. Saunders Company; Hutchinson Group Ltd; Leonard Lee Rue; Longman Group Ltd; International Union for Conservation of Nature and Natural Resources; W. Junk bv; Macmillan Publishing Co. Inc. for permission to reproduce Fig. 5.5 from *Ecology of population* by Arthur S. Boughey (© 1968 A. S. Boughey) and Figs 2.7, 2.22 and 3.19 from *Natural ecosystems* by W. B. Clapham Jr (© 1973 W. B. Clapham Jr); Hugh Maynard; the Editor, *New Scientist;* Oliver & Boyd Ltd and J. Tivy for permission to reproduce Fig. 56 from *Biogeography: a study of plants in the biosphere;* Parks Canada; the Director, Plant Breeding Institute; N. Pears; Princeton University Press; Reader's Digest Association Ltd (*Living world of animals*); Hans Reinhard; RIDA Photo Library; Bill Sanderson; *Scientific American* for permission to reproduce 3 illustrations from 'The plants and animals that nourish man' by Jack R. Harlan, copyright © 1976 by Scientific American Inc., all rights reserved; B. Seddon; R. V. Tait; T. Norman Tait; John Topham Picture Library; Fritz Vollmar; E. O. Wilson; S. Woodell. Figure 1.9 is reproduced by kind permission of the following: Harper & Row, Publishers, Inc. (Copyright © 1972 by Charles J. Krebs); R. T. T. Forman and Duke University Press (from R. T. T. Forman, *Ecological Monographs* **37**, pp 1–25. Copyright 1964, the Ecological Society).

Contents

PREFACE   *page* iii

ACKNOWLEDGEMENTS   iii

INTRODUCTION   v
What is biogeography?   v
Framework   vi

Chapter 1   DISTRIBUTIONAL PROCESSES   1
             1.1 Environmental factors in plant and animal
                  growth   4
             1.2 Community factors in distributions   7
             1.3 How can we characterise the distribution of
                  organisms?   10

Chapter 2   ECOSYSTEM PROCESSES   17
             2.1 Energy flow in ecosystems   17
             2.2 Mineral nutrients and their pathways   20
             2.3 Population dynamics   23
             2.4 Ecosystems in time   26
             2.5 Biological productivity   30

Chapter 3   BIOME PROCESSES   32
             3.1 Deserts   32
             3.2 The tundra   36
             3.3 Temperate grasslands   40
             3.4 Tropical savannas   42
             3.5 Sclerophyll ecosystems   46
             3.6 Boreal coniferous forests   47
             3.7 Temperate deciduous forests   49
             3.8 Tropical evergreen forests   53
             3.9 Islands   55
             3.10 The seas   56
             3.11 General   64

Chapter 4   MAN AND BIOGEOGRAPHICAL PROCESSES   65
             4.1 The different animal   65
             4.2 Domestication   67
             4.3 Simplification   74
             4.4 Obliteration   78
             4.5 Diversification   80
             4.6 Conservation   83

Chapter 5   ENVOI   90
             5.1 So what is biogeography?   90
             5.2 The real world   90

FURTHER READING   92

GLOSSARY   93

INDEX   96

Introduction

'There is no wealth but life'

(John Ruskin)

In many schools biogeography is one of the Cinderellas of geographical society. It is taught usually with physical geography and so suffers from juxtaposition with the considerably more popular geomorphology. Even when biology is part of the curriculum in the last two years of secondary school, the overlap and synthesis which are possible seem to happen rather rarely. One reason for the status of biogeography is that the teachers know relatively little about it, or at any rate do not value it as highly as, for example, geomorphology. This state of affairs probably reflects their own experiences as students, for biogeography was often relegated to a back seat in universities, polytechnics and colleges; only recent graduates of these institutions are likely to have found it ranked more or less equally with other branches of the discipline. Another possible reason for its unpopularity is perhaps that it is seen as uninteresting compared with competitors for limited school time. I find this difficult to believe: there is the intellectual challenge of mastering unfamiliar material, the contact with the ideas of eminent scientists past and present, the interest in hearing about faraway places, the opportunity to carry out field and laboratory work (and indeed the chance still to make original observations), and above all to study a field in which the systems of nature and the systems created by man meet. For biogeography cannot be studied realistically only as a branch of physical geography, since man has altered too many of the plants and animals and their habitats for it to be solely a 'natural' study. Neither, of course, can it be a totally 'social' study, for biological organisms do have their own lives outside human minds. So it can be an 'interface' subject with roots both in natural and social sciences but transcending both to become a very difficult but pioneering subject with, of course, a relevance to today's problems, and in this regard it is a microcosm of the discipline of geography where the whole is always something more than the sum of its parts. I hope something of the flavour of today's biogeography has found its way into this book: not only the excitement and the relevance but also (to be realistic) some of the hard concentration required to master basic material; like all intellectual achievement there has to be perspiration as well as inspiration.

What is biogeography?

It is not possible to give a simple definition of biogeography, for the word has been used in a number of different ways. Common to them all, perhaps, is the idea that biogeography is the study of the distribution of plants and animals over the surface of the Earth. This study has two main phases: description, in which we try to identify meaningful patterns in the distributions we observe; and explanation, in which we try to say how and why these patterns have come about. The attempt to identify patterns of distribution on a large scale (e.g. the world or a continent) is usually called **phyto-** or **zoogeography** (depending on whether it deals with plants or with animals) and these sciences together make up biogeography as it is understood by biologists. The study of pattern may take place also at a regional or local scale and the groups of organisms described are usually referred to as **communities,** their physical environment being called the **habitat.** Complementary to this view are investigations which attempt to link together plants and animals and their environment: to study the effect each has upon the other. Such an approach is essentially that of **ecology,** and the unit recognised is the **ecosystem,** which can be studied at any scale from a drop of water to the entire planet. This way of thinking is essentially functional, i.e. it emphasises how the systems work. The environment of plants and animals includes factors of a physical nature, such as climate, water and soil, but nowadays often comprises man as well. Our species has not only altered or removed many ecosystems but has changed the genetic material in many groups of plants and animals so that the outcome of their reproduction is more to our liking. These considerations have led one scholar to suggest that a geographer's approach to biogeography could be focused on the ideas (a) that man creates new **genotypes** (i.e. breeds plants and animals which perpetuate different characteristics from their ancestors), and (b) that he creates new ecosystems in all kinds of ways – by covering them over, by simplifying natural communities, by replacing wild species with tame ones and in many other ways. In order to assess the effect of our species under these two headings, it is essential to have information about conditions before we inter-

vene and so historical biogeography is clearly an important study.

Framework

We start by considering some of the fundamental processes which determine why plants and animals grow where they do, noting particularly that although we have to single out each process for description, in reality many processes are operating simultaneously. Some of this material will be familiar to those taking parallel studies in biology and they may feel able to skip the first pages and join the story at p. 18. This first section ends with the concept of an interactive system of plants, animals and their non-living environment – the ecosystem – and we go on to discuss the way in which an ecosystem functions at a local scale. So the flows of energy and matter, changes in the populations of organisms and the rate of production of living matter in an ecosystem are described. The ecosystem concept can be used at larger scales and so a major block of material deals with the major world formations of climate, soil, plant and animal life (e.g. forests, grasslands, the oceans) as they would be if they were in a natural condition; however, we have to note in passing that many of these formations (called **biomes**) have been modified by human activity. This in turn leads us to an examination of the main categories of human impact upon plants, animals and their ecosystems, using a simple classification of the processes involved. Lastly, there is a short section devoted to a discussion of the alternative future relationships of man and other living organisms. So far as is possible, each of the major sections is free standing, i.e. it can be read without the others as prerequisites; this lessens the possibility of intellectual progression but means that parts only of the book can be read if time does not permit detailed study of it all.

Chapter 1

Distributional Processes

'No pleasure endures unseasoned by variety'
(Publilius Syrus, 1st century BC)

Our starting point must be the enormous variety of life on Earth; the ability of living organisms to grow and reproduce in so many different habitats and to take such different forms. The basic classificatory unit of this variety is the **species,** which may be thought of conveniently as comprising all those individuals which can breed among themselves but which cannot breed at all freely with individuals from other groups. We currently recognise about 2 million species, and Table 1.1 gives some idea of the classification of different groups of living things. It is important to remember that it is unlikely for all members of one species to be identical and this characteristic of **variation** is important in the evolutionary change of plants and animals. Changes in the species present at various periods of Earth's history are shown by the fossil record, from which we know that some species are of quite recent origin, whereas others have had a very long span of tenure. We also know that many life forms have become extinct (the dinosaurs are the most obvious example), although the reasons for their demise are often far from clear. What is normally accepted is that constant change in the characteristics of species has been brought about by **natural selection.** By this we mean the way in which the environment of a plant or animal (including other organisms) exerts a pressure which ensures that the individuals most fitted to the environment

survive and reproduce. Hence the importance of variation (e.g. of size or behaviour) within a species (Fig. 1.1), for amongst the variety of individuals there will be some which are more likely to survive (especially in times of relatively rapid environmental change such as that of climate) than others and these are the ones which will reproduce. Their offspring may be different from their ancestors and so a drift in characteristics will occur; this may eventually result in a

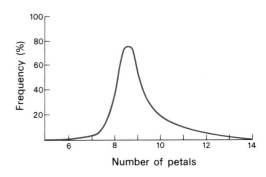

Figure 1.1 Every species of living organism shows some variation, provided sexual reproduction has taken place. Here, the number of petals in a sample of the lesser celandine (*Ranunculus ficaria*) exhibits a typical amount of variance; most individuals (about 80 per cent) have eight petals, but the range is from six to 14.

Table 1.1. Simplified schemes of plant and animal classification. Numbers are estimates of the number of species in each group.

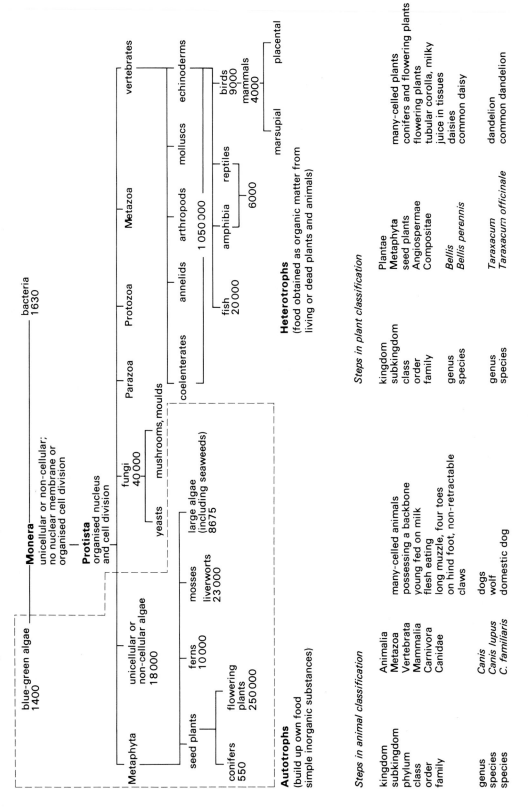

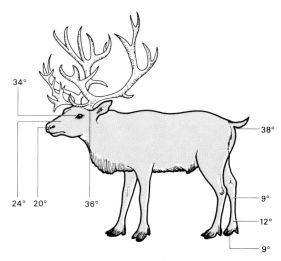

Figure 1.2 This sketch shows the temperatures on the body of a caribou when the air temperature around the animal is −30°C. The extremities of the body are able to tolerate quite low temperatures, but the fur and blood supply combine to keep parts such as the head distinctly warm.

parent group giving rise to reproductively isolated successors and so a new species has been formed. If a species is to avoid extinction, it must adapt to the prevailing conditions and there are many ways in which this has been achieved. The most obvious is **morphological** adaptation of gross order, such as the development of wings which enabled birds to occupy a range of habitats previously unoccupied by such large creatures, or the shape of fish which enables them to swim through water with the most economical use of energy. At a lower order of magnitude we might cite adaptations such as the webbed feet of aquatic animals, or the coats of Arctic animals which permit them to survive in very cold places (Fig. 1.2). More spectacular perhaps are the co-evolutionary adaptations, as for example where the reproductive parts of an orchid and the mouth-parts of an insect are developed to ensure both cross-pollination of the plant and a food source for the animal: the Late Spider Orchid, beautifully illustrated in the BBC book of *Life on Earth* (p. 84), is an example. But adaptation may not be so obvious: there are, for example, strains within plant species which show **physiological** differences in various environments. The mountain and high latitude plant *Oxyria digyna,* for example, comes out of its winter dormancy at different day-lengths in Arctic and mountain

Photograph 1   Two different plant species of the same genus. *Primula veris* and *P. elatior* (cowslip and oxlip respectively). The differences are not easily distinguished in these photographs but they are anatomically distinct, though clearly related closely. They are sufficiently close for hybrids to be formed when the two species grow together.

burrows where they can avoid over-exposure to the levels of solar radiation which would cause so much water loss that they would die. The movement of the whole organism from one habitat to another (migration) is another example of a behavioural adaptation: birds which fly long distances so as to remain in a temperate climate all year (e.g. by summering in the tundra and wintering in mid-latitudes) are another example. It is possible, too, for plants to exhibit behavioural adaptations, for some desert plants simply shed whole branches during periods of extreme drought.

The conclusions we must draw from this discussion of variety and adaptation are twofold. Firstly, that different species have different environmental **tolerances,** i.e. one species can grow and reproduce under conditions where another would be excluded (Fig. 1.3) – thus we have an uneven distribution of species over the surface of the Earth. Secondly, that even within one species, response to the environment is likely to be uneven because of the variation in individuals. There is therefore always the potential for evolutionary change. This adaptive capability may be absolutely essential to maintain life if the environmental conditions change.

1.1 Environmental factors in plant and animal growth

Many factors of the physical environment condition the distribution of an individual organism. Taken together they constitute the potential range of a species, i.e. all those places where it

Photograph 2 The flower of the fly orchid, *Ophrys insectifera.* This plant mimics the anatomy of an insect and cannot set seed unless visited by bees or flies which transfer pollen and cross fertilize the plants. If possible, compare the mimicry with late spider orchid in *Life on Earth*, p. 84.

habitats; Arctic plants have higher concentrations of **chlorophyll** and reach peak rates of physiological processes at lower temperatures than their mountain equivalents. But the plants of both habitats remain of the same species. Another method of adaptation is **behavioural:** some desert rats and mice, for example, are not very different in their anatomy and physiology from their temperate relatives but may behave differently. Many of them are nocturnal and spend the day in

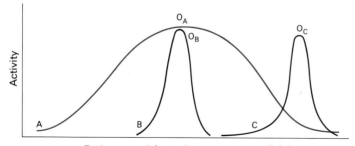

Environmental factor (e.g. temperature, light)

Figure 1.3 The patterns of tolerance for three species of living organisms are plotted on this graph. 'Activity' is a measure of successful adaptation to the environment, some measure of which (e.g. temperature or light) is plotted on the x axis. Species A has a wide tolerance, though best adapted at O_A – its optimum conditions. Species B has its optimum conditions (O_B) at the same value as O_A but its tolerances are also much more narrow. Species C is also a narrow-tolerance species, but its optimum at O_C is quite different from the other two. A and B, however, may well compete for resources around the values of their optima, but are unlikely to compete with C.

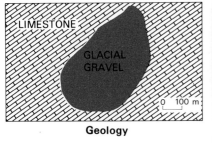 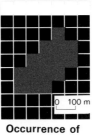

Geology **Occurrence of lime-demanding species in 100 m quadrants**

Figure 1.4 The distribution of an imaginary plant which is a calcicole, i.e. requires a high level of calcium-containing mineral nutrition. The geology on the left shows a patch of glacial gravel with an acid reaction overlying limestone with a calcareous reaction. On the right the presence or absence of the calcicole plant is recorded for every 100-metre square: it is virtually absent from the patch of glacial gravel.

might be potentially found. But this potential is rarely realised since the species' evolutionary history may, for example, be too recent for it to have dispersed into all the potential space; again, the environmental history of a locality may have ensured that the place is now beyond the limits of tolerance of a particular species. For instance, if glaciation has left a patch of gravel in a largely limestone area, then **calcicoles** (plants found in areas with a high calcium carbonate content in the soil, as above limestones for example) will probably be absent from the gravel area (Fig. 1.4).

If we look in more detail at contemporary environmental factors, then we must begin with the contribution of the Sun in providing *light* and *warmth*. Solar radiation of a particular wavelength is, of course, the basis of **photosynthesis** in which the radiant energy of the Sun is fixed as chemical energy in plant tissue – a process which is fundamental to most (but not all) forms of life. Not all the Sun's light is suitable for photosynthesis – only certain wavelengths – but in the course of a 'normal' day, most plants are saturated with light once the intensity of sunlight is above 10 000 lux (Fig. 1.5). The absence of light emphasises its importance: consider the very few species that grow on the floor of a forest with a dense canopy, or the filtering effect of water on light so that all the photosynthesis in water takes place in a zone less than 200 m deep. This is shallow compared with the depth of the oceans which are commonly 7000 m deep away from the continental shelves.

Warmth is another product of the Sun's radiation and is measured as a meteorological element and often expressed as a climatic pattern. Clearly, there are temperatures above and below which no organism can live at all (mice have lived and reproduced in a cold-store at –12°C; bacteria can be found at 90°C and their spores can survive

140–180°C), but most plants have an optimum temperature around 25°C and can function well between 10°C and 35°C. There may be critical times in life cycles: a certain temperature may be needed to ripen a seed or to end the **hibernation** of a pollinating insect and in certain years the critical value may not be reached. On a larger scale, it is notable that the northernmost limit of tree growth coincides more or less with the July isotherm for 10°C, and that reptiles are not found in the Arctic and relatively few species of this group are seen in temperate latitudes. In broad terms, the number of species per unit area diminishes away from the Tropics and towards the Poles (Table 1.2).

Since *water* is essential for life, its available quantity and seasonality (and sometimes its quality) also act as differentiating factors in plant and animal distribution. Under the wettest conditions there are the adaptations to an aquatic environment seen in several groups of animals,

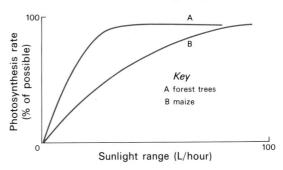

Figure 1.5 The photosynthetic rate as a percentage of the total photosynthesis possible (i.e. at 100 per cent the plant cannot carry out the process any faster), plotted against the quantity of incoming sunlight (L = lux), for (a) a forest tree and (b) maize. The forest tree reaches high levels of photosynthetic rate at a lower level of sunlight than the maize, thus producing much more organic matter per unit of time.

Table 1.2 General increase in species diversity with decreasing latitude in North America.

| | | Number of species: | | |
	Florida	Massa-chusetts	Labrador	Baffin Island
beetles	4000	2000	169	90
land snails	250	100	25	0
reptiles	107	21	5	0
amphibians	50	21	17	0
flowering plants	2500	1650	390	218

and water plants may possess adaptations such as an airtight compartment in a stem which then confers buoyancy on a leaf or flowering head. On land, the lowland equatorial climates are the main factors which have allowed the growth of luxuriant forests (pp. 53–5), the physiology of whose trees could not be maintained without a year-round input of rainfall. At the other end of the scale, lack of moisture produces adaptations in both plants and animals: the succulent habit of desert plants such as cacti is well known. Less obvious, perhaps, are modifications such as the ability to put down very deep roots: the root of a tamarisk shrub is reputed to have been traced some 30 m down during the excavation of the Suez Canal. Desert animals are frequently black (nobody quite knows why) and may possess the ability to store water, as in the case of the camel. Some arid-land plants, known as **ephemerals,** complete their whole life cycle from dormancy back again in a very short period after rains and an animal's breeding cycle may well result in the production of young at the time of maximum food availability after seasonal rains. Within such broad-scale considerations, moisture will produce different patterns of distribution of species, for example the lower limits of conifers on mountains in western North America seem to be affected by· the sensitivity to drought of particular species (Table 1.3).

Other factors of the physical environment which may place an overall limit on life, or which may produce a selective response from species or individuals, include gases such as oxygen and carbon dioxide. Since carbon dioxide is essential for photosynthesis, there can be little life without it. This gas is not likely to be severely limiting in the air but can be so in water since it enters the water mainly by mixing in the surface layer. Oxygen is essential for animal life and again may

Photograph 3 A perennial desert plant (the ocotillo cactus) of the Mojave Desert of California is surrounded by ephemeral plants which respond quickly to the winter rains and complete their life-cycle in a few weeks, from germination to setting seed.

be limiting in aquatic habitats in a sewage-contaminated river the bacteria may use up all the dissolved oxygen during their fast growth on the sewage, leaving very little for the fish which may then die (Fig. 1.6). But there are forms of life adapted to live without these gases, since we find

Table 1.3 **Lower limits of pine species on mountains in the USA.** Note. Since rainfall increases with latitude, the knobcone pine has become adapted to low precipitation levels and is not, therefore, sensitive to drought. The sugar pine normally grows where precipitation is highest and so is susceptible to long dry periods. The Coulter pine falls between the two.

	Lower limit (m)	Drought sensitivity
knobcone pine	850	least
Coulter pine	1200	intermediate
sugar pine	1600	most

bacteria which live in sulphur springs and also apparently in petroleum at depths of 4000 m.

We see later (pp. 20–3) that plants receive their combined mineral nutrition from the soil and so the quantity of these elements in a given soil may act as a differentiating factor. For example, soils which have lost most of their minerals due to **leaching** by soil water percolation will support only those plants which can tolerate very low supplies of the nutritive elements. In Britain, the growth of heather (*Calluna vulgaris*) on moorland soils is an example of this process at work. Soils derived from parent materials with a diversity of minerals and which have not subsequently lost them in runoff may support a much wider variety of flora. Soils which are very high in mineral salts, as in deserts or at the sea's edge, present special problems of growth to plants and those which are adapted to be salt tolerant are called **halophytes.**

Fire may be an element of the natural environment. Animals have to adapt to it by fleeing it or by migration in dry periods, but some plants have evolved a very thick bark which keeps the growing cells alive and these are called **pyrophytes.** Other plants have seeds which will only

Photograph 4   A sulphur spring at Yellowstone. Even in environments like this, bacteria can flourish: some species can use the inorganic chemicals as sources of energy.

germinate after they have been burnt and a number of pines have cones which only open and release the seeds after burning. These adaptations suggest the presence of fire over many millenia and are not found when fire appears to be a recent phenomenon due to human presence.

1.2 Community factors in distributions

It is not only factors of the physical environment which determine the presence and abundance of organisms, for the individuals interact among themselves: one of the most frequent relationships is **predation.** Mice may eat grass, for example, and in turn be eaten by owls. This may affect the growth of the grass and the numbers of mice and owls but is unlikely to account for the total absence of any of them unless the system gets out of balance. For instance, if the owls are exterminated, the mice become so abundant that they eat all the grass until they then all starve. But such predator–prey systems are usually finely tuned to avoid such extreme occurrences, just as a successful parasite may debilitate its host but will not kill it as this would deprive it of a home and sustenance (Fig. 1.7).

Competition is another form of community relationship and occurs between species (interspecific competition) and between individuals of the same species (intra-specific competition). Inter-specific competition may occur when, for example, two species may potentially occupy the same habitat (e.g. bracken fern and heather on a hillside) but where one factor, or a combination of

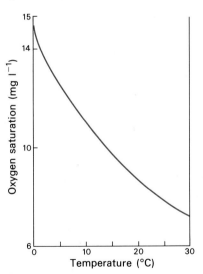

Figure 1.6   The oxygen saturation curve of water at different temperatures. The quantity of oxygen dissolved in water falls with rising temperature. The quantity of oxygen available, for example to fish, would thus drop off in summer in a temperate climate. If fast-growing organisms were then to compete for oxygen (e.g. bacteria from pollution) the fishes' chance of survival is much lower when the water is warm than when it is cold.

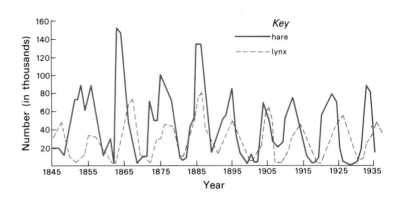

Figure 1.7 Changes in the abundance of lynx and snow-shoe hare, determined from skins received by Hudson's Bay Company. The peaks in numbers of the hares are followed, after a slight lag, by peaks in numbers of their predator – the lynx. However, the lynxes are never so efficient that they wipe out their food source altogether.

Photograph 5   An area of heathland in Sussex, England. The dominant vegetation is mostly heather but in the foreground and middle ground areas of bracken can be seen. The bracken and heather are competing in the same habitat. Apart from the contemporary dynamics of the vegetation (e.g. the colonisation by bracken), we need to remember that this was once woodland, relics of which can still be seen.

factors, may give one the edge over the other. The ability to withstand grazing by domestic animals such as sheep, the occurrence of fire, or the availability of nutrients, are examples. Intra-specific competition occurs in plants, for example, in germination when the emerging seedlings jostle for light, water and nutrients, with survival going to the most vigorous. In animals there is competition between males (including fighting) for the privilege of breeding, which seems to ensure that the 'best' genetic traits are passed to the next generation. Thus intra-specific competition is a potent force that exerts a strong

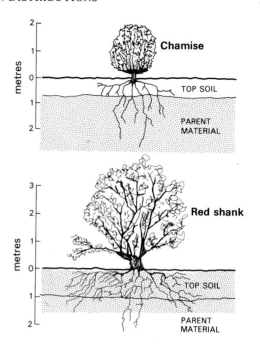

Figure 1.8 The separation of two closely growing desert species – chamise and red shank. The roots of chamise are sparse in the top soil, but penetrate some distance into the subsoil; those of red shank are nearly all in the top soil. Thus, although water is scarce, these two plants can grow closely together without competing for the small quantities of water that become available.

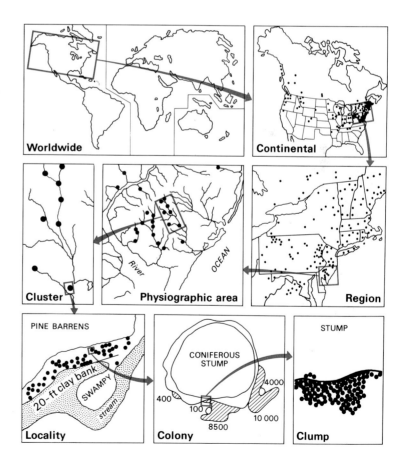

Figure 1.9 An American moss (*Tetraphis*) plotted at different levels of distribution, from the world (top left) to the clump (bottom right) which shows that it can only grow on the stable soils at the foot of conifer tree stumps. The more difficult question to investigate would be 'why is it only found in North America?'.

selective pressure on populations and it forces them in the direction of evolving differences that allow them to evade the competition for resources, i.e. to coexist rather than to compete (Fig. 1.8). Such selective processes are likely eventually to produce different species.

1.3 How can we characterise the distribution of organisms?

Given that an organism is unlikely to be evenly distributed in the world, for the various reasons sketched in above, it becomes essential to be able to characterise its distribution. Of considerable importance here is the scale at which we wish to work: Figure 1.9 shows very well how the distribution of a moss (*Tetraphis*) may be mapped on a number of scales: its world distribution is limited (for reasons unknown) to North America; within that continent, temperature and moisture provide the main constraints on its distribution. Locally, its moisture relations are highly important, as is its inability to grow on **unstable**

soils, so that it is confined to the presence of conifer tree stumps.

At a world scale, it is possible to recognise certain groups of plant species. There are, for example, groups which appear to have continuous worldwide distributions providing that suitable conditions are present: the palms are a tropical and subtropical example of such a group. Another example is given by a saxifrage species which is virtually continuous in the circumpolar zone but exhibits outliers in temperate mountains, where broadly similar conditions are to be found. Within the British Isles there are a few plants, e.g. the daisy (*Bellis perennis*) and the ribwort plantain (*Plantago lanceolata*) (Fig. 1.10), which exhibit a continuous distribution to the point of being found in every 1-km grid square, although obviously not necessarily continuously within that unit. On the other hand, there are numerous examples of discontinuous distributions where a plant is found in widely scattered places. Some examples are given in Figure 1.11 and the fascinating part is not so much the scatter

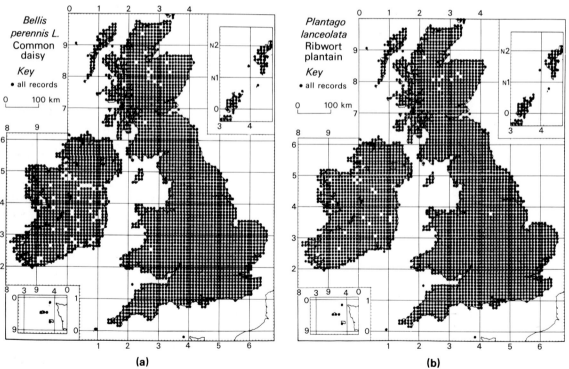

(a)   (b)

Figure 1.10 **The presence or absence of two plants, recorded for every 1-km grid square in the British Isles.** The plantain and the daisy are so common that they are found in almost every square. The blanks in Highland Scotland could be environmental, but what might be the cause of the single gap in the Midlands for the plantain? Are there really no daisies in lowland Gwent?

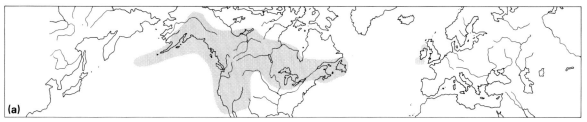

(a)

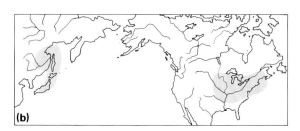

(b)

(c)

Figure 1.11 Some unusual worldwide plant distributions. At the top is an orchid (*Spiranthes romanzoffiana*) found commonly in North America and sporadically in the western British Isles. It has been suggested that seeds have travelled from West to East stuck to the feet of migratory birds. The middle map is of the skunk cabbage (*Symplocarpus foetidus*) in eastern North America and eastern Asia. Could climatic factors be responsible? Can we say anything unless we know the complete fossil record of the plant? The bottom map is of the members of the pitcher-plant family (the Sarraceniaceae). What might be the reasons for the outlier of distribution in South America? Is the range perhaps increasing, or might it in fact be decreasing?

pattern. But another explanation might be that the tree evolved in Africa and was taken to India and Australia by traders in early historic times. In the absence of fossil evidence of the tree in Australia and India, we might incline to the latter interpretation. Equally fascinating is the distribution of an orchid (see Fig. 1.11) which occurs in eastern North America and then again in the far west of the British Isles: here the distribution is probably related to the migration pattern of geese who occasionally carry viable seeds on their feet. The extreme cases of discontinuous distribution are plants which are exclusive to a particular area (**endemic**), because they have evolved there or because a barrier has prevented their spread: places surrounded by oceans, by extreme climates or by mountain ranges are often rich in such species, as are mountain ranges which are 'islands' of a particular sort. Thus 45 per cent of the species of the flora of the Canary Islands are endemic, 82 per cent of the Hawaiian flora and 85 per cent of that of St Helena.

Some workers have attempted to go beyond the distribution of individual species and to talk of floral regions – an area of any size distinguished by an assemblage of certain types of plants, a high proportion (50–70 per cent) of which are endemic to it, as well as possessing many others which occur largely within its boundaries. Such floral regions can be built up into a hierarchy with a characterisation of the world at the top (Fig. 1.13),

but the derivation of explanations for the pattern. Sometimes fossil distributions from either the Quaternary or earlier geological eras will point to possibilities but mechanisms are often dificult to find. The case of the baobab tree is an interesting one (Fig. 1.12). This plant is found in widely scattered parts of the world and we might be led to the explanation that the plant evolved on an ancestral continent which has broken up and so **continental drift** is the major cause of the present

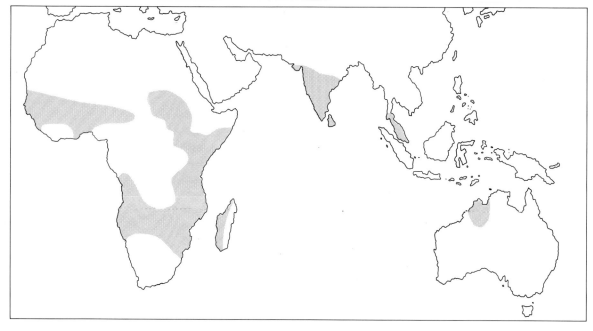

Figure 1.12   The genus *Adansonia* **(baobabs) around the Indian Ocean and in Africa.** Could the break-up of a primeval continent in far-off geological time have caused this disjunct pattern? Or might traders have carried the tree from Africa eastwards during the historic period?

but when compiled are clearly subject to such an aggregation of factors that the patterns lack any explanatory power save the most general, such as world climate – a criterion which can be levelled also at the biomes which are discussed in Chapter 4.

The terms used for floral distribution can be used also for animals. At one level there are endemics confined to small areas: the islands of the Galapagos each have a different **subspecies** (Fig. 1.14) of giant land tortoise, derived from a common ancestor. (These differences and other features of the Galapagos fauna are credited with being very important to Darwin in his elaboration

Photograph 6   A specimen of the Australian baobab, *Adansonia gregorii,* a genus whose curious distribution is shown in Figure 1.12. The large trunk is most likely an adaptation for the storage of water over a long dry season.

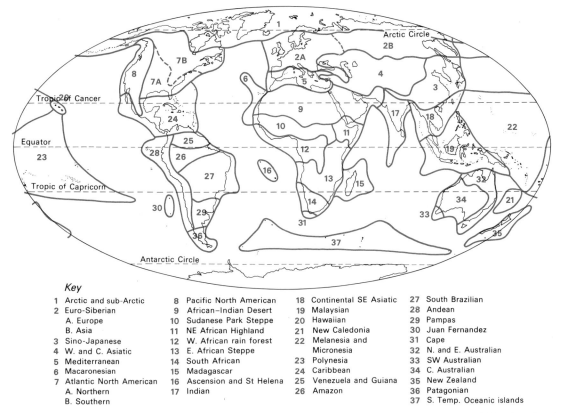

Key

1	Arctic and sub-Arctic	8 Pacific North American	18 Continental SE Asiatic	27 South Brazilian

1	Arctic and sub-Arctic
2	Euro-Siberian
	A. Europe
	B. Asia
3	Sino-Japanese
4	W. and C. Asiatic
5	Mediterranean
6	Macaronesian
7	Atlantic North American
	A. Northern
	B. Southern
8	Pacific North American
9	African–Indian Desert
10	Sudanese Park Steppe
11	NE African Highland
12	W. African rain forest
13	E. African Steppe
14	South African
15	Madagascar
16	Ascension and St Helena
17	Indian
18	Continental SE Asiatic
19	Malaysian
20	Hawaiian
21	New Caledonian
22	Melanesia and Micronesia
23	Polynesia
24	Caribbean
25	Venezuela and Guiana
26	Amazon
27	South Brazilian
28	Andean
29	Pampas
30	Juan Fernandez
31	Cape
32	N. and E. Australian
33	SW Australian
34	C. Australian
35	New Zealand
36	Patagonian
37	S. Temp. Oceanic islands

Figure 1.13   An attempt to amalgamate many maps of plant distribution into a world map of 'floral regions' in which a certain proportion of the plants are found only within that region. Apart from being descriptive of a pattern, does such a map suggest anything explanatory?

of a theory of evolution.) Less spectacularly perhaps, but of equal scientific importance, is the fact that on the Fijian islands, the ranges of the species of doves (genus *Ptilinopus)* do not overlap: no island has two species of this genus. Back at the subspecific level, the remote Hebridean island of St Kilda has its own endemic subspecies of house-mouse.

The distribution of an animal at a continental scale may reflect many factors such as food supply, climate and evolutionary history. The two beetles in Figure 1.15 exemplify different ranges for Europe: one appears to be a reasonably cold-tolerant animal (but not arctic-alpine), whereas the second is clearly intolerant of very cold winters although by no means a subtropical animal. It is remarkable that, although the overall ranges of these two beetles are so different, they converge into an almost identical distribution within Britain.

At a world scale we can recognise distinct zoo-geographical regions (**realms**) within which the fauna shows particular characteristics (Fig. 1.16). For example, the tropical part of the Ethiopian realm has many species of animals but this diversity diminishes to both south and north. Some species of shrews, the aardvark, the hippo and some monkeys, are entirely confined to this realm. In the Palearctic realm there is a diverse fauna but its variety is less than in the Oriental and Ethiopian realms: there are few reptiles, for example, and no monkeys. Perhaps the most distinct fauna is found in the Australasian realm, for here are found not only primitive mammals which lay eggs, e.g. the ant-eaters and the duck-billed platypus, but a particular richness of bats, birds and reptiles. The most distinctive element is the **marsupial mammals,** which have occupied most of the habitats filled by 'ordinary' mammals elsewhere in the world; there is thus a grassland herbivore such as the kangaroo, equivalent to the bison and llama of the Americas and the

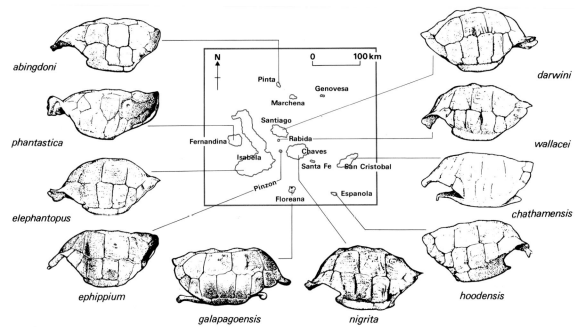

Figure 1.14 These shells of giant land tortoise (*Geochelone elephantopus*) of the Galapagos Islands each come from a different island in the group. Isolation has produced these differences from one common ancestral stock; now they are regarded as subspecies, for they do not interbreed – being land tortoises.

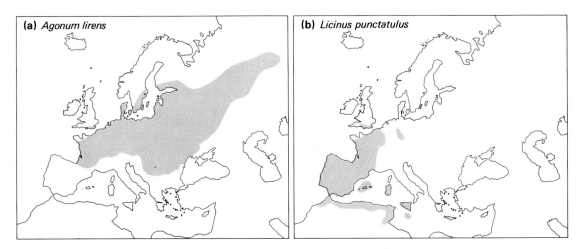

Figure 1.15 **The distribution of two European ground-beetles.** *Agonum lirens* (a) is largely a creature of central and northern Europe, whereas *Licinus punctatulus* (b) is a much more southern animal, stretching to the Mediterranean littoral. However, interestingly, their distributions coincide in southeast England.

antelopes of Africa and Asia and a nocturnal desert marsupial mouse equivalent to the kangaroo rat of the Sahara.

However, species do not exist by themselves in nature: even maps which tell us a lot about the characteristics of an organism, such as the cold-tolerance of arctic-alpines, say nothing about its neighbouring plants, the soil it grows in, what, if anything, eats it and what happens to it when it is dead. All these (and other) factors can be studied only at the local scale, in terms of the assemblage of species within a given habitat, an assemblage called the community. Its minimum property is the simultaneous presence of several species in

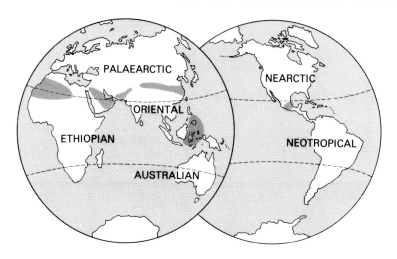

1.16 The world's zoogeographic realms. The shaded areas are transitional between adjacent areas. The divisions are made on the basis of distinct faunal assemblages – for example, marsupial mammals are confined to the Australasian realm; the Nearctic realm lacks monkeys; and only the Ethiopian and Oriental realms now have elephants.

the same area, but it may also exhibit a certain stability, tending back towards a particular condition if it has been disturbed, as a gap in a forest will eventually be recolonised with trees. A community can be studied in terms of a number of attributes: there is first the question of its **species diversity** – what species of plants and animals live in it? Secondly we can investigate its **physical structure**. If it is a terrestrial community, then we

can distinguish layers in many plant communities: trees, shrubs, herbs and mosses, for example. The relative abundance of the various species may be measured and this concept is closely related to that of **dominance** (Fig. 1.17). Not all the species in the community, after all, are equally important in determining its nature, only a few exert a controlling influence because of their size, numbers or behaviour. These **dominants**

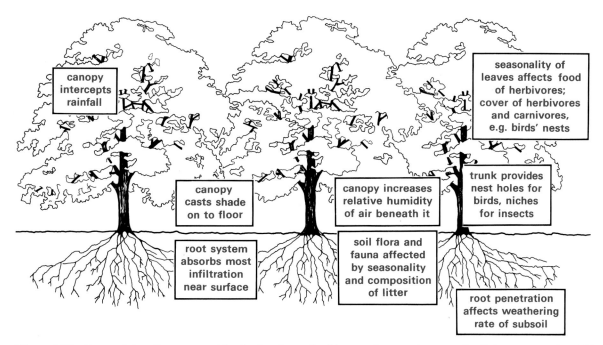

Figure 1.17   Some of the effects exerted by the dominant (in this case the trees in a woodland) in a community. All kinds of flows of energy and matter are controlled by the dominants when living. An interesting exercise would be to imagine the trees removed and the different comments that could be made in each box.

control the conditions under which associated species must live to a considerable extent: thus in a forest the trees are dominant; on a fern-covered hillside the bracken is dominant.

Lastly, we must investigate the linkages between species: 'who eats what?' is perhaps the simplest way of putting it; **food chain** and **food web** are also familiar terms. This leads to a very important way of looking at communities and their interactions not only within themselves but also with their habitat. In this way we can study the flows of energy and matter which constitute the metabolism of the community and enable us to compare very different kinds of community in the same terms. The study of plants and animals interacting with each other and with their non-living environment is called ecology, and a spatial unit of interacting organisms and their physical habitat is called an ecosystem.

Chapter 2

Ecosystem Processes

'The golden apples of the sun'

(W. B. Yeats)

The idea of a continual exchange of matter and energy between living organisms and their surroundings is at the heart of the idea of the ecosystem (Fig. 2.1). Thus an American ecologist, E. P. Odum, has put forward a formal definition of an ecosystem as:

> Any unit that includes all of the organisms in a given area interacting with the physical environment so that a flow of energy leads to . . . an exchange of materials between living and non-living parts within the system . . .

It seems clear that the study of ecosystems should start with a discussion of the flows of energy within them. We need, therefore, to consider the way in which mineral nutrients enter the system and are combined with the energy to make living **organic matter.** This matter manifests itself in the form of populations of individual species and we shall consider how these grow and are controlled. We shall consider how ecosystems change in time as they progress from an initial colonisation of a biologically dead area up to a stable self-renewing maturity. The organic matter accumulation of the mature community over time can be measured and this rate of production of organic matter (**biological productivity**) is an important aspect of ecosystems on the world scale.

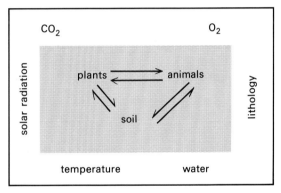

Figure 2.1 A simple diagram of an ecosystem concept. In the inner box the plants, animals and soil all interact with each other by exchanging matter and energy. They also interact with their surrounding environment, so that this boundary is drawn as an intermittent line. The diagram could represent a very small ecosystem, such as a patch of lawn, or the whole planet.

2.1 Energy flow in ecosystems

We start this discussion with green plants. They contain the pigment chlorophyll and can fix energy from the Sun into organic molecules, i.e. molecules containing carbon. This process is called photosynthesis and is the source of all the energy in living things and thus in all ecosystems, with the exception of a few lower organisms which can fix energy in the absence of light. The importance of photosynthesis for life (and hence for man's economies) cannot be over-estimated.

The process occurs in two steps: the first uses light energy absorbed by chlorophyll to split a water molecule, releasing oxygen; the second, which does not require light, uses the energy in several steps to reduce carbon dioxide to **carbohydrates.** This can be summarised in the equation:

$$6CO_2 + 12H_2O \xrightarrow[\text{(chlorophyll)}]{\substack{\text{Solar energy} \\ 2816 \text{ kJ}}} C_6H_{12}O_6 + 6O_2 + 6H_2O$$

$$\underset{\substack{\text{carbon} \\ \text{dioxide}}}{} \quad \underset{\text{water}}{} \qquad\qquad \underset{\text{sugar}}{\phantom{C_6H_{12}O_6}} \quad \underset{\text{oxygen}}{} \quad \underset{\text{water}}{}$$

where kJ = kilojoules (1×10^3 Joules).

The carbohydrate $C_6H_{12}O_6$ can be converted by the plant into starch and stored; it can be combined with other sugar molecules to make cellulose which is a basic structural material in plants; or it can be combined with elements such as nitrogen, phosphorus and sulphur to produce **proteins, nucleic acids** and all the other constituents of living cells. Some of the sugar produced by photosynthesis is used as an energy source by the plants themselves for growth, the maintenance of tissues and biochemical processes. This process is called **respiration** and can be summarised as:

$$C_6H_{12}O_6 + 6O_2 \longrightarrow 6CO_2 + 6H_2O + 2830 \text{ kJ}$$

$$\underset{\text{sugar}}{\phantom{C_6H_{12}O_6}} \quad \underset{\text{oxygen}}{} \qquad \underset{\substack{\text{carbon} \\ \text{dioxide}}}{} \quad \underset{\text{water}}{} \quad \underset{\text{energy}}{}$$

It is worth noting that this energy is converted to heat in the course of its use by the plant and so is never available for use again within the ecosystem since it is a dispersed low-grade heat energy rather than the concentrated high-grade chemical energy which is incorporated into plant tissues. The amount of energy which appears as a net increment of organic matter is therefore dependent upon the balance between the rates of photosynthesis and respiration. Put simply and symbolically:

$$N = P - R$$

where P = gross production in plants, or the total energy assimilated by the organism in a given time; R = respiration, that part of the assimilated energy converted by the plant to heat or mechanical energy or used in life processes; N = net production, or the increase in organic matter or total energy content in a given time. It appears as an increase in the 'living weight' of plants (and by extension it can be used of the whole ecosystem), i.e. an increase in **biomass.** The rate of accumulation of biomass (i.e. weight of living matter/unit area/unit time) is called **net primary productivity** (NPP).

Looked at in terms of a world scale, about 520×10^{22} joules (J) of energy strike the top of the Earth's atmosphere every year, of which about 100×10^{22} J reaches the Earth's surface and is of a suitable wavelength for photosynthesis. Of this, some 40 per cent is reflected back into space by deserts, snow, ice and oceans, leaving 60×10^{22} J as the 'pool' for photosynthesis. Of this, a large quantity is respired and the remainder appears as biomass, i.e. as accumulated organic matter. The total amount of biomass produced annually by green plants is estimated to contain about 170×10^{19} J (much of it from **phytoplankton** in the oceans) and so the average utilisation of the

Figure 2.2 The flow or flux of energy from the Sun to living plants. Box 1 shows the incident energy at the top of the atmosphere. A small proportion only (box 2) of this arrives at the Earth's surface which is of suitable wavelength for photosynthesis, and about 40 per cent of this is reflected back into the atmosphere from the oceans and deserts (box 3). Box 4 thus forms the 'pool' for photosynthesis (P) but much is used (5) in respiration (R). Box 6 represents the actual biomass – the living plant tissue. Its value is about 0.2 per cent of box 1. Less than 1 per cent of the quantity in box 6 becomes food consumed by man.

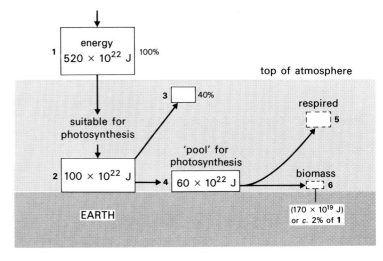

energy present in light of the right wavelength by the flora of the globe is about 0·2 per cent. Less than 1 per cent of that 0·2 per cent is consumed by man as food (Fig. 2.2).

The fate of energy which appears as net primary productivity gives us a way of constructing a model of an ecosystem. In its simplest form (Fig. 2.3) it consists of a series of boxes which represent stored energy, stored as organisms, i.e. biomass, together with lines which indicate flows between the boxes, and so a model of a simple energy transfer system – **food chain** – is constructed. Normally, some of the plant material continues as storage in perennial plants, some is eaten by **herbivorous** animals, and some dies to form a litter on the soil surface. Herbivorous animals are eaten by **carnivores** or else join the litter upon death. Carnivores have only the litter layer to look forward to unless they in turn are predated upon by yet another carnivore. Organisms which live in the litter layers immediately start to break down the litter, which is then chemically decomposed by the soil flora and fauna or **decomposer** organisms. The type of litter and its specialist organisms play a key role in the recycling of mineral nutrients in the system. Each stage in the chain is called a **trophic level.** The plants are the first trophic level (**producers**), the herbivorous animals are the second trophic level (primary **consumers**) and the carnivorous animals are the third trophic level (secondary consumers). In most terrestrial ecosystems three or four trophic levels are found. Because there are usually several different species of plants and animals at each trophic level, each animal having its own

feeding patterns, the idea of a **food web** rather than a food chain is more realistic. The web itself may be complicated by the fact that not all species feed constantly at one trophic level. Some may move seasonally between trophic levels, or may be herbivorous at one stage of their life history and carnivorous at another (e.g. tadpoles and frogs), or may simply be able to feed opportunistically as a source becomes available. But the concept of the web allows us to see that a multiplicity of species at any one trophic level might result in a greater stability in the system should one be removed (Fig. 2.4).

In their energy relations, all such chains and webs are subject to one of the basic laws of physics governing energy flow. Put simply, this law requires that at every transfer step in the ecosystem some energy will be degraded from a highly concentrated chemical form to a highly dispersed form as heat which cannot be recycled into chemical energy but must be radiated out to the atmosphere and then to space and so lost to the ecosystem. Thus at each trophic level a conversion to heat takes place, which means that less energy becomes biomass at the succeeding trophic level, especially at the herbivore and carnivore levels. A carnivore may have to expend a great deal of energy catching its prey and many carnivores are thus adapted to eating at relatively infrequent intervals, as are, for example, lions and snakes. One result of this law is that in a given ecosystem the amount of energy available as food to the next trophic level is reduced at levels successively away from the plant producers: if the energy content (usually given as the **calorific**

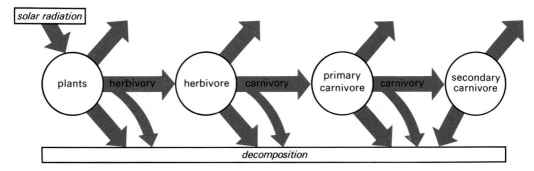

Figure 2.3   Energy flow through a food chain. At each stage some energy is available as food for the next trophic level, some is lost as excretory products, some as decay of dead organisms and some as respiration. Thus the quantity of energy decreases down the chain away from the plants; the number of secondary carnivores in a given area is thus likely to be much smaller than that of the plants. Much of the energy 'lost' to the chain ends up with the decomposer organisms which in terrestrial ecosystems are in the top few centimetres of the soil.

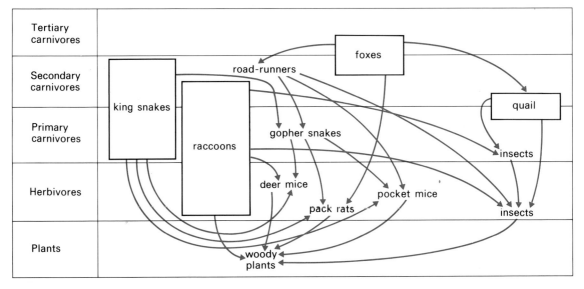

Figure 2.4   A simplified food web, showing some of the more complex linkages in an actual ecosystem, in this case the chaparral of California. The various trophic levels demarcate the animals of different groups and it is noticeable that not all animals are confined to one trophic level – racoons, for example, are both carnivores and herbivores, and the food of foxes comes from a variety of trophic levels. However, they are all ultimately dependent on the plants.

value) of each trophic level is plotted, a pyramidal shape is obtained (Fig. 2.5). A simple grass field ⟶ mice and voles ⟶ hawks and owls ecosystem clearly follows the normal trends in this respect: the absolute number of predators is quite small compared with the number of grass plants. The sizes of individual animals may increase up the chain since it may well be an advantage for a predator to be larger than its prey. Some ecosystems may appear to have more energy at their consumer stages than in their plant producers: this can be so because organic matter is imported across the ecosystem boundaries, e.g. carried by running water into a pond, or by the tides into an estuary. These facts have obvious relevance for the human use of living things. If an ecosystem (whether natural or man-directed) is being cropped for food, there will be much more energy available per unit area if man eats as a herbivore than as a carnivore, assuming that the plants yield as much edible matter as the animals (Table 2.1).

2.2 Mineral nutrients and their pathways

Apart from energy, life is sustained by a number of chemical elements which enter ecosystems via the plants. These elements come from the materials of the crust and atmosphere and, if involved in the growth and maintenance of organisms, they may conveniently be termed **mineral nutrients.** The mineral nutrients needed for life are dominated by four elements – carbon, hydrogen, oxygen and nitrogen – and indeed these elements, combined in various ways, represent all but a tiny portion of the Earth's terrestrial vegetation and hence most of the living matter on the planet. There are 12 other elements which have been found to be essential to life – potassium,

Table 2.1   Food for man as herbivore and carnivore. Note that the potatoes represent 30 per cent of the synthesised organic matter, but that the beef is only 4 per cent. However, remember that we do not eat beef for energy, it is primarily a luxury way of getting animal protein (1 Mcal = 1000 calories = 4190 J).

	Potato crop	Grassland
total organic matter synthesised per acre (Mcal)	24 750	28 000
net human food after preparation	7500 (potatoes)	1500 (beef)

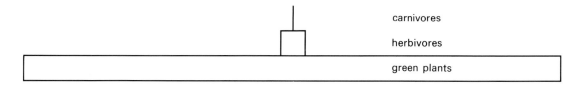

Figure 2.5 An ecosystem with three trophic levels, arranged as a pyramid of energy content (which would be measured in joules/area) showing that the amount of energy as organic tissues at each trophic level decreases away from the plants, because of the losses set out in Figure 2.3.

calcium, magnesium, phosphorus, sulphur, iron, copper, manganese, zinc, molybdenum, boron and chlorine. Oxygen, carbon and hydrogen form the basic cell structures and are major components of fats and carbohydrates. To them is added nitrogen for the synthesis of proteins; the inclusion of phosphorus allows the building of nucleic acids and **cytoplasm** and helps with the transfer of energy through cells. Sulphur is necessary for the formation of the amino acids, which are the building blocks of proteins; calcium helps to strengthen cell walls and without magnesium chlorophyll production would be impaired. Many of the other elements act as catalysts in the biochemistry of living organisms. The latter thirteen are usually derived from the bedrock, whereas the first four come ultimately from the atmosphere. Thus neither soil nor atmosphere alone can support life: interaction between them nearly always occurs. Only oxygen and hydrogen are freely available in large quantities, since they are atmospheric gases in plentiful supply, and so a characteristic feature of the pathways of the other elements is that they circulate between living organisms and non-living (**abiotic**) 'pools' of various scales. These pathways are generally cyclic: they involve the use and re-use of the atoms of the nutrients and since they also include a non-living phase, they are usually called **biogeochemical cycles** (Fig. 2.6).

At a local scale, a principal result of an ecosystem's development through time seems to be the creation of 'tight' or 'closed' nutrient pathways which recycle essential nutrients within the system. Any losses are small and are balanced by inputs from the non-living parts of the ecosystem (Fig. 2.7). The two major pathways for recycling nutrients are a return, firstly by way of animal excretion and secondly, by way of microbial

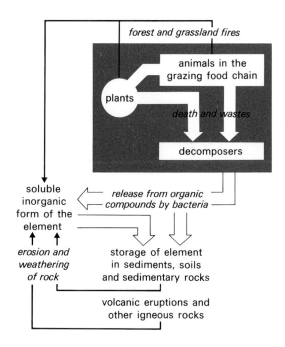

Figure 2.6 A generalised biogeochemical cycle for most chemical elements. The shaded area includes the phase when the elements are present in living organisms, from which they are released by fires and by bacterial action in the soil (or other substrate in aquatic environments). The storage phases in sediments, soils and sedimentary rocks may vary considerably in time, from perhaps a few weeks in some tropical forest soils to millions of years in accumulating sediments on the sea bed.

decomposition, of plant and animal detritus. Both function in ecosystems, but the first dominates in systems where most energy transfer follows a chain through a herbivore and then several carnivores, as happens when the producer stage consists of plankton; the second is dominant

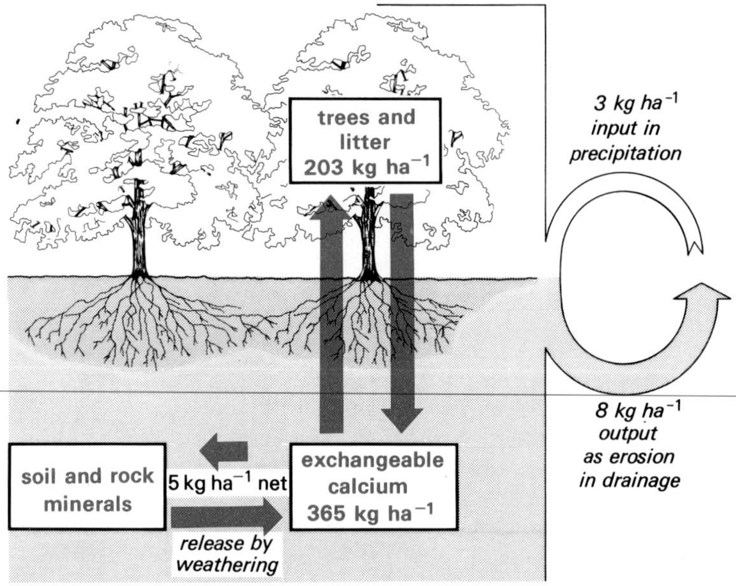

Figure 2.7 The biogeochemical cycle for calcium in an undisturbed deciduous forest in New Hampshire, USA. This is a 'tight' nutrient cycle and so the quantities of calcium circulating inside the free-soil system are much greater than those lost to it by erosion and runoff. However, these losses are balanced by inputs from precipitation and from the weathering of rocks.

in, for example, temperate forests and grasslands where more energy is transferred via the litter or detritus chain than goes through the above-ground herbivore–carnivore route. All these processes need energy for their metabolism and this is one way in which the energy content of an ecosystem is 'used'. An ecosystem which is to be relatively stable through time must, therefore, have adequate storage of both energy and nutrients to guard against periods when their supply is cut off. Photosynthetically active energy cannot be stored for plants to use, but for all the other trophic levels energy-rich organic matter can be stored as part of living plants or as animals, for example. Equally, dead organic matter can be stored as litter.

Mineral nutrients are often stored as part of this organic matter as well. A mechanism must be available to release them and in the case of litter the role of the decomposer organisms is especially important. The residence time of elements in the litter of the ecosystem is, however, usually short compared with the biomass, so that the loss by runoff in terrestrial systems is minimised. For example, the turnover time of nitrogen in the biomass of a temperate deciduous forest was 88 years, whereas in the litter layer it was 5 years.

The plants and animals of the decomposer layer do not normally enter geographical thinking, but it can be seen that their importance is immense. The decomposer organisms consist of two main groups: the **microflora** (mainly bacteria and fungi) and the **invertebrates,** including such animals as earthworms, beetles and springtails, mites and various larvae. Dung beetles and termites may be among the more visible elements of the fauna, especially in the Tropics. Roughly speaking, the invertebrates comminute the debris physically by feeding off the tougher elements in it, and the smaller pieces are then subject to decomposition by fungi and bacteria. The invertebrates themselves are, of course, subject to predation and parasitism and so the litter layer forms an ecosystem in microcosm, except that it is fuelled by dead matter and not directly by the Sun (Fig. 2.8).

The role of the vegetation in holding minerals in system cycles is especially marked in the Tropics where abundant rainfall would quickly leach minerals from soils and weathered rocks. One of the mechanisms hypothesised is the transfer of minerals from quickly decomposing litter directly to the uptaking roots via a coat of fungi which selectively take up the minerals required by the tree. Certainly, at any one time most of the nutrients are in the biomass and the mineral soil has a very low nutrient content compared with a deciduous forest in temperate latitudes.

Another feature of tropical forest vegetation is that the cycling time of mineral nutrients is short; the turnover time for minerals seems to rise with latitude, which presumably exerts its effects

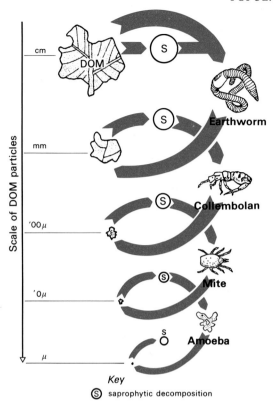

Key

Ⓢ saprophytic decomposition

Figure 2.8 The gradual degradation of dead organic matter (DOM) in the litter layer of an ecosystem. The scale shows how the actual size of litter particles is reduced (1 μm = 10^{-3} mm) by various organisms all adapted to feed on different size particles. The process is aided by the presence at each level of saprophytes (S), which are mostly fungi that obtain nourishment from the DOM and in turn prepare it for further breakdown.

through the length of the growing season and the activity season for the soil flora and fauna. A world pattern of different cycling times, treated to the receipt of solar energy, seems to emerge.

The biogeochemical cycle concept may also be used in habitats disturbed by man. The uptake of nutrients such as nitrogen and phosphorus by trees has led to suggestions that forests might be the most effective sewage treatment works yet discovered, since they might deal with the sludge as if it were a litter layer and absorb nitrogen and phosphorus as well, which even three-stage treatment may not do at present.

As with energy flow, the models based on biogeochemical cycles can tell us a great deal about the functioning of an ecosystem. We do need to recall that these cycles are not separate from

energy flow but that an ecosystem has to use part of its energy in maintaining them, as distinct from adding to biomass, for example. The living tissues made possible by mineral nutrients are also the chemical storage places of the energy in the concentrated form in which it can be utilised by other organisms. The studies of energy flow and mineral cycling must therefore proceed together.

2.3 Population dynamics

The flows of energy and mineral nutrients through an ecosystem manifest themselves as actual animals and actual plants of a particular species. Groups of the same classificatory unit which exist within definable limits of space and time are called populations. Some characteristics of interest are the absolute size of the population (which can be very small – the Javan rhino may have a population of less than 50 individuals – or which can be very large – we can only guess at the number of bacteria in the world), and its density, i.e. its relation to the volume of space in the habitat.

The main biological criteria that affect population size and density in a given area can be summed up in graphical form as:

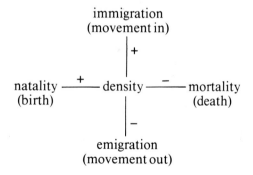

This scheme can be applied at any scale and to any classificatory group including our own. The concepts of immigration and emigration need little explanation, except perhaps to stress the difficulty in measuring them that exists in real ecosystems; and that, as with the ecosystem concept itself, the placing of boundaries across which an individual becomes part of one population rather than another can be problematical.

The factors which actually determine population size are very complex. In plants, competition between individuals of the same species is often a

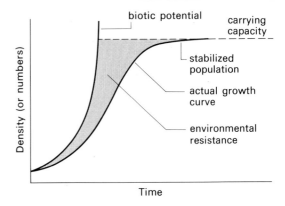

Figure 2.9   A graph of the growth of the population of an organism with time. After an initial phase of slow growth, there is generally a period of fast growth until some upper limit is reached (the 'carrying capacity') at which the population becomes stabilised. Were it not for external factors ('environmental resistance') the reproductive potential ('biotic potential') of the population would carry it to levels far beyond those represented for the carrying capacity on this diagram. (Without the environmental resistance, a single bacterium dividing into two every half-hour would cover the Earth's surface a metre deep in bacteria within a few weeks.)

Photograph 7   Browse-lines. (a) A red deer (*Cervus elephas*) reaching up to browse on the lower branches of a tree, and (b) the outcome of browsing by cattle and ponies in the New Forest of England, with a near horizontal line to the base of the foliage of the background tree (photo by Keith Barber).

regulator of density; indeed some plants grow better in a mixture than in single-species stands. Predation by herbivores may also exert a strong effect upon density, e.g. St John's Wort (*Hypericum perforatum*) is a weed of grazing land in many parts of the world, displacing forage palatable to stock and even poisoning domesticated animals if eaten in quantity. In California, its density has been reduced to about one-hundredth (i.e. 1 per cent) of its former level by the introduction of leaf-beetles of the genus *Chrysolina*. That larger herbivores may also have

an effect can often be seen by the erection of **exclosures** which keep them out of various types of vegetation, or when a herbivore itself undergoes a population decline. In Europe, for example, the decline of rabbit populations following the spread of myxomatosis allowed many grasses and woody species that would formerly have been eaten to flourish and species unpalatable to the rabbit, such as ragwort (*Senecio jacobea*), diminished in abundance in the face of the new competition. With the return of high densities of rabbits, ragwort became an effective competitor once more.

The regulation of animal populations is similarly complex, and short-cycle fluctuations of considerable amplitude are quite common in short-lived and rapidly breeding groups. Nevertheless, over a period and given a stable environment, most groups preserve characteristic levels of abundance so that we can think of a species as 'rare' or 'common'. There may be fluctuations in abundance but if the habitat remains stable, the populations will regain their characteristic level. As abundance increases, the population is generally affected by various influences which bring about the stabilisation of its numbers and density. The simplest form of this effect is increased mortality due to starvation since there is insufficient food. Predation may increase: as lemming numbers rise in the tundra, so their predators are said to immigrate from adjacent areas. Alternatively, at higher densities the proportion of individuals affected by a parasite may be increased because of easier transmission of the parasite from host to host. All such effects, where the regulatory factor operates more strongly at high densities than low, are called **density-dependent factors.** Some of these are very closely interrelated, as with the oscillation of

predator and prey numbers – the predator regulates the numbers of its prey and in turn is itself regulated by the prey numbers. A concept which emerges is that of **carrying capacity** (Fig. 2.9). This is the total number of individuals of a species that will live in an ecosystem under certain conditions and, although it clearly applies to plants, is generally a term used of animal populations. The carrying capacity is achieved when the regulatory processes (**feedback**) acting on the population density bring it into line with the available resources for the species after a period of rapid growth. These resources are

Photograph 8   Sea-birds on the Bass Rock in the Firth of Forth. It is clear that the population is most likely restricted by the availability of nesting sites than by factors such as the supply of food.

inevitably finite and one or more of them will prove limiting for a species. The limiting resources may be constant in time and space, like space itself. If only this is limiting, then the carrying capacity is the number of organisms that can be packed into the space – for example, barnacles on a rock or the number of nesting sites for seabirds on a steep cliff. It is more likely that the limiting resources may be variable over space and time and an equilibrium has to be reached that can accommodate the variations in their supply. This means that a population may be able to 'over-exploit' a resource at a time of its particular abundance but that it will fall back when the supply diminishes. Once a limiting factor has been identified in an ecosystem, it may be possible for a human manager to manipulate the system to raise it. In the California chaparral (a vegetation type familiar to all of us from cowboy movies and car chases in the suburbs of Los Angeles) fire can be used to keep down the proportion of woody vegetation and increase the leafy vegetation within the reach of deer. The carrying capacity of burnt chaparral for deer becomes much higher than unburnt areas, not only because there is more browse but because the leafy material is higher in protein than woody vegetation.

A population dynamic of considerable interest at present is that of the human species. After a long period of very slow growth and with definite checks and recessions, the human population entered on a rapid phase of growth in about the 17th century AD, with a growth rate in recent years of 2 per cent per annum, which means that the absolute size of the population doubles at 35-year intervals (Fig. 2.10). We would expect that an equilibrium level will be achieved at the carrying capacity of the planet for our species, but two highly contentious questions emerge – how is that carrying capacity to be defined and how is the equilibrium to be achieved? The former is not merely biological and nutritional, but subject to cultural and aesthetic preferences as well; the latter could be simply biological (war, famine and starvation, disease, spontaneous abortion from stress, low-level violence) but presumably it can be cultural also through the means of population-control programmes.

2.4 Ecosystems in time

Ecosystems change over time: they may start from zero after an event such as the eruption of a volcanic cone from the sea (Fig. 2.11); and they change on smaller scales as when an old tree in a forest is windthrown and replaced by a series of herbs, shrubs and trees until a mature forest is

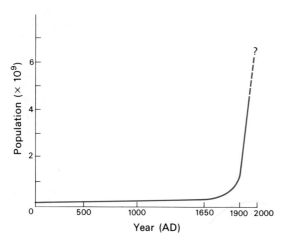

Figure 2.10 An outline curve of the growth of the human population through time ($1 \times 10^9 = 1000$ million). Where and when will the stabilised population of Figure 2.9 be reached? What forms does the environmental resistance take now? What forms might it take in a future of yet higher absolute numbers and densities?

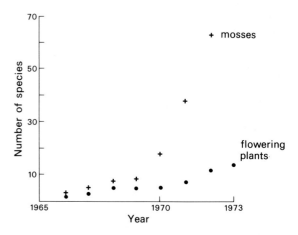

Figure 2.11 Colonisation and succession on the newly created volcanic island of Surtsey, off Iceland. The crosses represent the number of species of mosses, the dots that of flowering plants. Clearly, the mosses are able to adapt more readily to the conditions of bare rock, but a steady increment of flowering plants is present as well.

present once again. The progression from the establishment of life at a particular place up to an ecosystem which is stable and self-renewing is called **succession** and the mature stage is often referred to as the **climax.** Where this climax lasts for hundreds of years and appears to be in equilibrium with environmental conditions, it is sometimes called the climatic climax. Some apparently mature ecosystems are so old that we do not know their original stages of succession: some parts of the tropical rain forests, for instance, may have been in much the same state for 100 million years. Other ecosystems are much newer: the forests of temperate lands started as bare land after their last glaciation and proceeded through successional stages of lichen and moss tundra, scrubby birch and aspen, through pine forest to a stable mature forest of oaks, limes, elms and beeches. A pond left after the decay of ice gradually fills in as the succession of plants moves inwards, filling it with organic debris. Forest may then colonise or in some places acid bog populated by plants nourished only from rain-water may become established. At each stage of succession a different dominant may emerge.

The mature ecosystem has several distinct features. The food chains are usually long and they interweave in a complex way, whereas at

Photograph 9 Plant communities and their habitats. The vegetation exhibits zoning, with dry-land tree vegetation, marginal reed-swamp and open-water plants. At the same time, these plant communities are undergoing succession, since the pond is contracting so that the zones are gradually moving inwards. In a relatively short time, if there is no human intervention, the pond will disappear and its site will be covered with woodland.

Photograph 10   A mosaic of ecosystem types. The first is a virtually inert area, with no visible evidence of life: an area of lava of recent origin on the island of San Salvador in the Galapagos. The second picture is an early stage of succession, when some higher plants have become established among volcanic ash, on the Teide National Park in Tenerife. The third shows a mature natural forest capable of regenerating itself: a stable and mature ecosystem (the Waipoua State Forest on the North Island of New Zealand).

earlier stages of succession the chains are simpler and more linear. The biomass is higher than at earlier stages and represents, we assume, the 'best' that can be achieved under the prevailing conditions. Species diversity will be higher than in early stages of succession and the ecosystem will probably be well stratified into layers. A mature forest, for example, will have a tree layer, a shrub layer (which includes immature forest trees) and a ground layer of grasses, herbs and mosses. Tropical rain forests typically have three tree layers but no others; boreal coniferous forests have a tree layer and a ground layer only. The nutrient cycles will be effectively closed and local, with little loss to the outside, in contrast with early successional ecosystems. Lastly, the mature

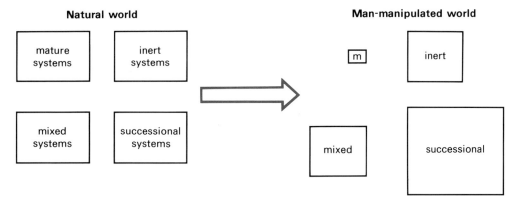

Figure 2.12   The world's ecosystem mosaic. On the left are the different kinds of ecosystem in a natural world largely unaffected by man (for explanation of terms see the text); on the right are the relative proportions of the different systems after millenia of human impact. The mature systems are much diminished in area but the successional (including systems such as agriculture) are greatly enlarged, as are inert areas where there is no life.

system will be self-renewing and have a good resistance to disturbance from outside – it will repair itself.

The natural world before the significant influence of man consisted of a mosaic of ecosystems at different stages. There were large areas of mature ecosystems, smaller areas of successional systems (some filling gaps caused by catastrophes in the mature systems, some colonising newly available areas such as recently deglaciated terrain, weathered lava flows or **eustatically emergent** coasts), some mixed mature and successional areas at zones of contact where a small climatic change is critical for the ecosystem type, and some areas without living organisms, such as volcanoes, ice caps and shifting sand. Man has altered this mosaic, increasing the inert and the successional at the expense of the mature (Fig. 2.12). Later, we must consider the question 'does this matter?'

2.5 Biological productivity

Another way in which the energy and matter present in living organisms are made manifest is in the rate of production of organic matter. As explained earlier, the fundamental process is the production of organic tissues incorporating solar energy by green plants, and its most important practical expression is net primary productivity (NPP) which is the material actually available for harvest by animals and for decomposition by the soil fauna and flora or their aquatic equivalents. Net primary productivity therefore combines phenomena such as the non-uniform distribution of incident radiant energy and the variable conditions of moisture supply, with living features such as the genetic properties of the plants which are the primary producers. Net primary productivity is usually measured as dry organic matter synthesised per unit area per unit time and expressed either as g m^{-2} yr^{-1} or kg ha^{-1} yr^{-1}. It can also be expressed as the calorific value of the dry organic matter, in kilocalories or joules (1 cal = 4.2 J). At any one time, the standing crop of living organisms present per unit is the biomass.

Recent research has given us an accurate idea of the NPP of the major ecosystems of the world and the data are set out in Table 2.2. Looking at these data we can see immediately that since the tropical rain forests have a year-round growing season and a high biomass, they would perhaps be expected to produce the greatest amount of

Table 2.2 NPP for the world, c. 1950 (source: Whittaker & Likens 1975).

Ecosystem type	Area (10^6 km^2)	Mean NPP (g/m^{-2}/yr^{-1})	Total production (10^9 t/yr^{-1})
tropical rainforest	17·0	2200	37·4
tropical seasonal forest	7·5	1600	12·0
temperate forest: evergreen	5·0	1300	6·5
temperate forest: deciduous	7·0	1200	8·4
boreal forest	12·0	800	9·6
woodland and shrubland	8·5	700	6·0
savanna	15·0	900	13·5
temperate grassland	9·0	600	5·4
tundra and alpine	8·0	140	1·1
desert and semidesert scrub	18·0	90	1·6
extreme desert: rock, sand, ice	24·0	3	0·07
cultivated land	14·0	650	9·1
swamp and marsh	2·0	3000	6·0
lake and stream	2·0	400	0·8
Total continental	149	782	117·5
open ocean	332·0	125	41·5
upwelling zones	0·4	500	0·2
continental shelf	26·6	360	9·6
algae beds and reefs	0·6	2500	1·6
estuaries (excl. marsh)	1·4	1500	2·1
Total marine	361	155	55·0
World total	510	336	172·5

organic matter in the course of a year. Not so widely appreciated is that limited parts of the oceans, such as estuaries and coral reefs, reach the NPP of tropical forests, although for the whole globe they are outweighed in absolute terms by the immense areas of open ocean whose NPP is more like that of the tundra. Tropical grasslands overlap with some of the woodlands of less favourable climates, and there is a big gap between these ecosystems and those of tundra, desert scrub and desert. The position of agricultural land is of some interest: at a world average of 650 g m^{-2} yr^{-1}, it exceeds the average figure for grasslands but falls well below most of the forests. The column for total production in

Table 2.2 emphasises the role of the forests in providing the bulk of the NPP of the world (62·8 per cent of the continental area, 42·8 per cent of the total), whereas cultivated lands produce only about 7·7 per cent of the terrestrial and freshwater NPP.

Because of the energy flow characteristics of ecosystems, the **secondary productivity** of these areas (i.e. the animal biomass per unit time) is much smaller than that of the plants. In the tundra the biomass of plants exceeds that of the animals by 15 times, in deciduous forests by 300 times and in tropical forests by 500 times. But we must not forget the often important role of the decomposer organisms whose productivity does not feature in NPP data. In a temperate grassland, some 13 per cent of the energy in the grasses went to herbivores but 86 per cent went to the decomposers directly: in several ecosystems there may be as much life below the soil surface as above it.

If it is now estimated with reasonable accuracy that total world NPP is about 170×10^9 t yr^{-1}, of which 50–60 t yr^{-1} is from the oceans, does this have a significance for our use of **biota** as a resource, bearing in mind, for example, the low proportion of world NPP contributed by agriculture (7·7 per cent of terrestrial NPP), and the low productivity of much of the oceans? First of all, it puts in perspective the activities of man. The energetic magnitude of world primary production is estimated at $28\,900 \times 10^{17}$ kJ yr^{-1} whereas man's use of fossil fuels and other industrial energy in 1970 was $19\,600 \times 10^{16}$ kJ yr^{-1} – $c.$ 7 per cent of NPP. The concern is that the latter figure has been doubling every 10 years, whereas the former may diminish as a result of human impacts, and that the impact of the industrial energy is not uniform. We need to know which high natural productivities are being affected by it: estuaries are one example. Man's harvest of food also looks small compared with biosphere production – about 0·72 per cent of the energy of global NPP. This does not mean to say that increasing food production is likely to be an easy task, but it does show that there must be some potential for higher production for all purposes.

Chapter 3

Biome Processes

The most extensive ecosystem unit which it is convenient to designate is called the biome. This consists of a dominant life-form (e.g. deciduous trees, grasses) which extends over an area that corresponds to a particular distribution of soil types and climate. Thus, although the species of tree or grass may vary from continent to continent, it is possible to recognise the deciduous forest or grassland biome in each. Thus the biomes of the world are major world-scale regions which integrate a number of factors into an intuitively recognisable whole – deserts, forests, savannas, oceans, etc. The map of biomes (Fig. 3.1) is basically a vegetation map (and in some atlases is called just that) and conceals the fact that there are soils and animals associated with each biome type, but the very patterns will suggest that world climate is perhaps the dominant factor in producing the mosaic.

The map is not one of contemporary reality, for man's activities have transformed many of the areas plotted on it. It is rather a map of what the world would be like if human activities suddenly ceased and all the resulting successions were telescoped in time. Much of the scientific work on biomes has been done on present-day relicts (as in the case of European deciduous woodland, for example) and we cannot be sure that the ecology of these remnants does not differ from its condition in the original state.

We shall consider the major biomes in turn, starting with the terrestrial low-productivity systems and moving up the productivity scale to the tropical forests; then we shall do the reverse for the aquatic systems. Terrestrial and aquatic systems will be bridged by an account of islands.

3.1 Deserts

These biomes are found in areas of the world where the average precipitation is below 250 mm yr^{-1} and is often irregular; dew fall during cold nights is the only other source of water. Much of the precipitation is evaporated rapidly by the high levels of solar radiation. The biotic response is firstly in scantiness of vegetation so that there is usually more bare ground than plant cover, and secondly in marked adaptations of both plants and animals to enable them to survive long periods of drought or the lack of access to free water. The aridity is usually compounded by very high daytime temperatures so that evaporation greatly exceeds precipitation in the soils and the potential for water loss from individual organisms is very high. For example, the ground temperature may reach 60°C during the early afternoon, when the air temperature is around 40°C. At the same time the relative humidity is at its lowest, perhaps lower than 20 per cent, having fallen from a peak of 60 per cent at 06.00 h.

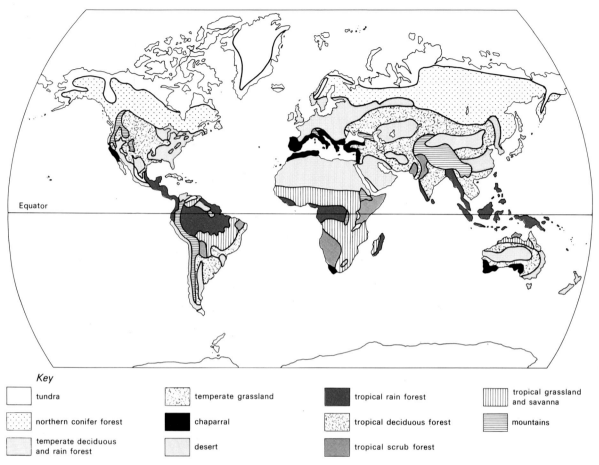

Figure 3.1   A map of the major biomes of the world as if they were not affected by human activity. In reality, large parts of several of them (e.g. temperate grassland, temperate deciduous forest) have been substantially modified. The biomes are known by their dominant plant formations, but possess distinct animal communities and soil characteristics as well. Clearly, there is a strong element of climatic control in their distribution.

Deserts are found in a number of topographic–climatic situations. Subtropical deserts are the results of semipermanent belts of high pressure in the Tropics, from which air is warmed by compression on descent and becomes very hot and dry; the Sahara is an example of this type. Cold currents offshore may produce cool coastal deserts such as the Namib, the Atacama and the coastal desert of Baja California. The other type of cool desert occurs in the interior of great continental masses, such as the Gobi Desert north of the Himalayan massif which helps to keep precipitation away just as in the rain-shadow deserts such as those of the Great Basin and Mojave (southwestern USA) and Chihuahua (northern Mexico), which are in the lee of mountain ranges such as the Sierra Nevada, the San Gabriel Mountains and the Sierra Madre Occidental.

The deserts are thus one of the most unfavourable environments for life on Earth, as are regions of permanent ice and snow and their margins and the ocean depths. We need not be surprised that NPP is so low (of the order of 90 g m^{-2} yr^{-1} but as low as 3 g m^{-2} yr^{-1} in some rocky and sandy places; compared with temperate deciduous forests at 800 g m^{-2} yr^{-1}) and that much of it is underground (Fig. 3.2). In many desert ecosystems about 80 per cent of the NPP is underground away from the shrivelling rays of the Sun, and of the above-ground biomass perhaps only 1 per cent will be green since photosynthesising tissues are also potential areas of water loss.

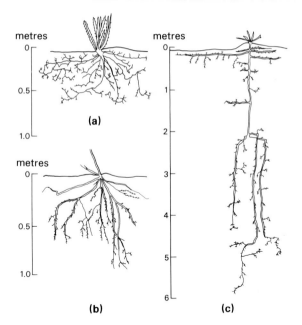

(a)

(b)

(c)

Figure 3.2   Rooting systems as adaptations to drought in desert plants. (a) is a succulent plant which can store a great deal of water and so can rely on near-surface roots collecting rain from sporadic precipitation events. (b) is a grass which can store less water and so must have a more reliable supply from the subsoil – hence the deeper and spreading root system. (c) is an *Acacia* bush which adopts a dual strategy of having immediately subsurface roots to collect any moisture sporadically available, together with a very deep rooting system stretching down to permanently available water in the subsoil and rock.

Not surprisingly, the soils of the deserts are low in organic matter.

Plants exhibit special adaptive mechanisms for survival in these conditions. Spreading root systems are one such mechanism: although plants such as the saguaro cactus (*Carnegia gigantea*) appear to be widely scattered, the lateral roots may be in contact with each other. Succulent plants such as cacti store water in roots, stems or leaves, to the point where some may be used as emergency water sources by desert dwellers and travellers. To prevent water loss by transpiration, plants show a variety of adaptations such as thick waxy **cuticles,** downy hair coverings, small fleshy leaves or rolled-up leaves reduced to virtual spikes, and the ability to shed leaves, twigs, whole branches or they may even stop growing altogether when water is particularly scarce (Fig. 3.3). There is also the ephemeral habit in which annual plants remain dormant as seeds until rain comes. The water dissolves an inhibiting chemical in the seed coat and germination takes place followed by the entire life cycle of the plant through to setting seed within a few weeks.

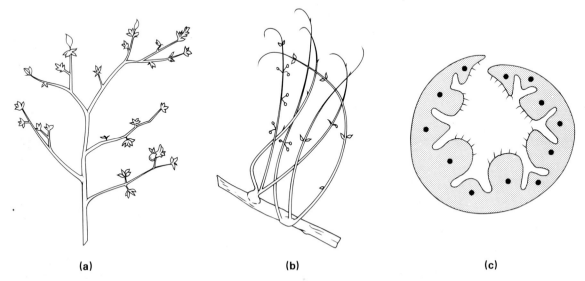

(a)   **(b)**   **(c)**

Figure 3.3   Some above-ground adaptations of desert plants. (a) and (b) show very small leaves in relation to the size (*c.* 25 cm) of branch. In (a) parts of the branches may be shed in extreme droughts. (c) is a section (2 cm diameter) of the leaf of a desert grass; its rolled nature helps to reduce water loss from the inner surface, as do the hairs which reduce air movement.

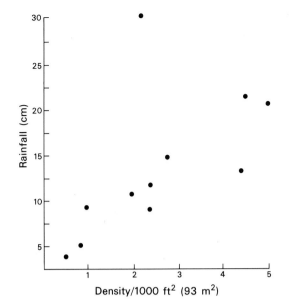

Rainfall (cm)

Density/1000 ft² (93 m²)

Figure 3.4 A scatter diagram showing the relationship between average annual rainfall and the density of the desert saltbrush *Larrea divaricata* in the Mojave desert of California. A very close relationship between higher precipitation and plant density can be inferred from the diagram.

The outcome in terms of plant communities is a vegetation closely related to water; as more water becomes available, the density of plants increases (Fig. 3.4). Relatively moist areas, such as the Mojave, have a vegetation of low (0·5–1·0 m) shrubs (such as creosote bush set 2–3 m apart) as the 'dominants' with other shrubs interspersed. At intervals, clumps of low (0·2–1·0 m) succulents such as *Opuntia* are found. When the rains come (regularly in spring in this case), the ephemerals grow and the desert blooms. At the other extreme, the dune lands of the Sahara have sparse populations of only a few plants, such as the drinn grass (*Aristida pungens*), and in the rocky and stony deserts there may be large areas with no vegetation at all except where

Photograph 11   A desert biome scene in Algeria. The shifting dunes of the background are bare of vegetation but, in the foreground, a more stable substrate has allowed the growth of a few shrubs which are adapted to the extremely arid conditions.

a little finer soil material can accumulate and support a grass or a small shrub.

Animal life is by no means absent: a number of adaptations have evolved which facilitate the survival of the creatures in the prevailing conditions of heat and drought. Most species are nocturnal, since they minimise water loss by being active at night and probably avoid some potential predators as well. Any animal which has a moist or porous skin, such as the desert forms of worms and slugs, is always nocturnal. Some animals are active in the day but even they mostly leave the ground surface by burrowing, climbing plant stems or flying when the temperature reaches 50°C; only a few grasshoppers, beetles and spiders are active during the very hottest conditions.

Physiological adaptations are numerous and mostly centre around a lack of need for free-water intake and around mechanisms for avoiding its loss. Many birds, insects and rodents, for example, gain their water in their food rather than from free moisture; when water becomes available, the African and Asian wild asses (*Equus asinus* and *E. hemionus*) can drink it in quantities amounting to a quarter of their body weight. The camel can lose 30 per cent of its body weight but can then drink up to 120 litres of water at a time. Water loss is minimised by a variety of means, of which perhaps one of the most common in vertebrates is an excretory system which produces concentrated urine and relatively dry faeces, and the animal may have very few sweat glands. The antelope ground squirrel (*Citellus leucurus*) can drink salty water more concentrated than that of the sea, and the sand rat (*Psammomys obesus*) secretes a urine up to four times as concentrated as sea water. By contrast, the North American jackrabbit does not burrow and ingests no free water, but it is suggested that its very large ears radiate heat to the air when the animal is resting in the shade.

As with plants, a flush of rain will transform the apparent barrenness into an abundance of life. Insects rapidly move in to pollinate flowers, dung beetles appear and roll away the droppings of mammals and the ants harvest grass seeds. Caterpillars, crickets, flies and wasps, rarely seen at other times of the year, also descend on the ephemeral plants. The rains may trigger off other processes as well: the maturation of the desert locust (*Schistocerca gregaria*) occurs in response to aromatic chemicals produced by desert shrubs

at the time of the rains. Calving of mammals usually occurs at the time of the rains if these are regular: at the rainy season, the camel ruts and then has a 12-month gestation period so that young are produced at a time of maximum forage.

Even though the deserts are sparsely inhabited, any human activities which affect the above-ground plant biomass have the potential to change the ecosystem radically. Grazing of domesticated animals is the obvious method and so it need not surprise us that extension of deserts is taking place where heavy grazing of domesticated animals occurs – part of the phenomenon of **desertification**. Nevertheless, pastoralism of animals such as the camel can be a stable systems in deserts provided that movement is frequent. In deserts used for recreation (as in North America) severe damage to the ecosystem can result from vehicles which damage the plants, especially succulents, and, for instance, the use of dead cactus 'skeletons' for fires robs the soils of organic matter as well as destroying the scarce cover for arthropods such as insects and spiders. The natural rate of decomposition of leaves, animal shoots and seasonally dying parts appears to be fast, but woody material is much slower to decay. This means also that human-introduced inorganic debris remains virtually unaffected except by mechanical weathering for a very long time. So although the deserts scarcely look like a fragile biome, they are certainly capable of modification by man.

3.2 The tundra

This biome is found mostly in the northern hemisphere where it is the most northerly formation, since beyond it lies either the Arctic Ocean or permanent snow and ice. An analogous set of ecosystems is found on sub-Antarctic islands and the fringes of the Antarctic continent.

In the northern hemisphere, the boreal forest (see p. 47) gives way northwards to a zone of scattered clumps of trees in a low open vegetation and then finally to a broad belt of treeless vegetation in which shrubby willows, birches and alders are the highest vegetation. The transition appears to coincide with a mean daily temperature of at least 10°C in the warmest month, but where there is a growing season of more than 3 months. In both North America and Eurasia the boundary between forest and tundra lies further north in the west where climate is moderated by relatively warm westerly winds and a warm offshore current. The ground, however, remains frozen all year (**permafrost**), except for the few top centimetres which thaw during the summer, and water which is frozen for the rest of the year is then available to the plants. Precipitation is low (250 mm yr⁻¹) but is accompanied by slow rates of evaporation even in summer. Low-lying areas, therefore, tend to be waterlogged, whereas ridges are very dry because even snow is blown off them by the winds. Thus denuded, the soil freezes to a lower temperature than where an insulating cover of snow is found. In central Alaska, an air temperature above the snow of $-50°C$ was measured at the same time as $-11°C$ was recorded at the surface of the soil beneath the snow.

The growing season is not so short that higher plants such as herbs, grasses and low shrubs cannot grow, and is extended by the long hours of daylight during the summer. An NPP average of 140 g m⁻² yr⁻¹ is achieved, which is low but higher than deserts and the open ocean (see Table 2.2). With higher plants, much of this productivity is below ground; for example the ratio of root : shoot NPP at Point Barrow, Alaska, was 1 : 1·2 (a temperate deciduous forest ratio would be in the order of 1 : 3) and the litter layer acts as a considerable store of nutrients because of the very slow rates of decomposition. Areas of relatively high vegetation ('low tundra') are characterised, therefore, by a thick spongy mat of living and partly decayed vegetation. In summer the surface is saturated with water and small ponds are scattered throughout. The plants tolerant of these conditions are predominantly grass and sedge – for example, near Point Barrow the coastal plain is dominated by the grass *Dupontia fischeri* and the sedges *Carex aquatilis* and *Eriophorum scheuzcheri*. In sheltered valleys, thickets of dwarf birches (*Betula nana*) and creeping arctic willows

Photograph 12   A herd of caribou *(Rangifer tarandus)* **grazing on tundra vegetation in Alaska.** Even though the picture is of a sheltered valley, none of the vegetation approaches more than shrub height.

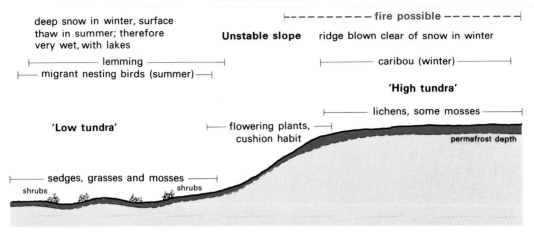

Figure 3.5   A sketch section showing the relationships of different types of tundra, with indications of some of the intensities of the various processes forming the ecosystems.

(e.g. *Salix arctica*) can be seen. In drier places, heathy plants such as *Cassiope tetragona* (bell heather) are found. Most plants are less than 20 cm high and thus lie beneath the snow in winter. The less productive 'high tundra' (with low vegetation) is much lower in moisture and frequently lacks a permanent snow cover in winter. The soils are low in organic matter and are vegetated by only a sparse growth of mosses and lichens, with only the latter in the most exposed places. Such vegetation is, however, permanently available for herbivores to eat during the winter (Fig. 3.5).

The animal life is conspicuously adapted to the climate in one way or another. Warm-blooded animals which live above the snow are generally covered in warmth-retaining fur (e.g. the musk-ox, *Ovibus moschatus;* Arctic hare, *Lepus arcticus;* wolf, *Canis lupus;* caribou and reindeer, *Rangifer tarandus*). There are other anatomical and behavioural adaptations: the snow-shoe hare (*Lepus americanus*) has pads on its hind feet twice the size of those of an ordinary hare to enable it to cross snow; cleft-footed animals which sink into the snow will make sunken trails, and they will clear away snow with their front feet to get at food beneath. They will be accompanied often by birds such as ptarmigans which also feed on the exposed vegetation. Migration is also practised, with the yearly round of the caribou herds (which are responding to the availability of food), a much-studied feature. The caribou drop their fawns on the tundra during May, having wintered at the forest edge. They then range up to the shores of the Arctic Ocean, frequenting high

ground to avoid the flies. They are constantly attended by wolves who keep the herds moving, which helps to avoid over-use of the forage (Fig. 3.6). During the summer there is a brief burst of invertebrate life, in which insects are most conspicuous, especially those of the mosquito and blackfly kind. The tundra is also a favoured nesting place for many species of migratory birds, few of which winter in the north. They arrive in spring and carry out their reproductive cycles very rapidly, so that the young are ready for the southward migration at the end of the short summer. Animals which live beneath the snow, such as the lemming, have relatively thin fur and possess few of the modifications required by the animals of the open air of winter.

Some animal populations in the tundra are marked by oscillations of abundance. The snow-shoe hare, for example, fluctuates on a 9–10-year cycle, with a population peak being followed by a crash. Its chief predator, the lynx, follows the same cycle of abundance with a 1-year lag. Perhaps the most famous cycle of abundance is that of the lemming, where the genera *Lemmus* and *Dicrostonyx* have species which in both Eurasia and North America become very abundant and then achieve spectacularly high mortality rates. At the especially high peaks which occur at 4-year intervals, the famous lemming migration occurs in Northern Europe, when the populations migrate southwards irrespective of any obstacles in their path. In North America, the ecological effect of the peak lemming numbers on the vegetation was shown in experimental plots, where ungrazed areas had 36

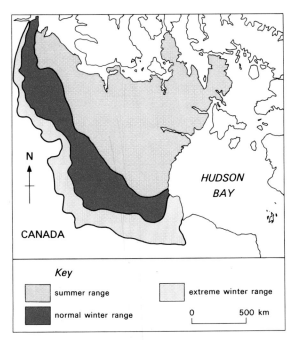

N

HUDSON
BAY

CANADA

Key

summer range   extreme winter range

normal winter range   0   500 km

Figure 3.6   The ranges of the caribou in the Canadian North. Area 1 is the summer range where the animals calve and try to escape the flies, and graze on vegetation which is snow-covered in winter. Area 2 is their normal winter range at the junction of the tundra and boreal coniferous forest. Area 3 is their range in an extremely bad winter, where they are driven south into the coniferous forests which, for this species, are not an optimal source of food.

per cent more grass in August than those utilised by lemmings. One effect, however, of heavy grazing by small mammals and invertebrates is that more solar radiation can reach the soil. Thus, temperatures rise and the soil flora and fauna are more active and so nutrient cycling is, to some extent, accelerated.

Much more stable are those animal populations which are linked to the food webs of the sea. Seals, and to a lesser extent walrus, are obvious examples, together with various species of sea birds, many of which (such as skuas and gulls) can feed off either sea or land. At the top of this chain is the polar bear (*Thalarctos maritimus*), now a threatened species and the subject of a circumpolar international convention.

The impact of man in such environments is bound to be strong, for with such a short growing season plants have low recuperative and colonising powers. Antarctica is a specially protected area, but the Arctic has a long history of human habitation by hunter–fishers – the Inuit or Eskimo in North America and similar groups in Eurasia – and a shorter but much more manipulative occupance by Western cultures. The hunter–fisher cultures were always low in density and lived almost exclusively off animals such as seals, caribou and fish, supplemented seasonally with berries, birds' eggs and chicks and small whales. But the Arctic is now exploited for oil, gas, minerals, tourism and for military purposes and the impacts on the tundra can be profound, not only by industrially based users but modernised natives as well. Road construction, fire and oil spills all increase the depth of thaw and the amount of surface subsidence. Such effects, in general, become less severe in the High Arctic because there is so little vegetation that its damage exerts less influence on the soil. Increased human presence has affected the population of land animals: it is said that seismic testing using explosives depletes a large area of Arctic foxes, and accidental fire on the dry lichens of exposed ridges leaves them unvegetated for decades and robs herbivores of a source of winter food. Many of man's influences have come together in their effect on the barren-ground caribou (*Rangifer tarandus arcticus*) of Canada, which declined from several million before European contact to 672 000 in 1949 and 200 000 in 1958. Excessive hunting was one cause, especially in the years 1949–60 when very low numbers of calves were born. It is difficult to enforce catch limits because of the terrain and because many of the hunters are Treaty Indians who cannot legally be obliged to observe hunting regulations. Also, the exploitation of the Canadian North has caused many forest fires in the lichen-rich forests where the caribou winter and so their food supply has been diminished.

Even the distant islands of the sub-Antarctic have not escaped human influence. Dogs, cats and rats were introduced to Macquarie Island (54°S), and rabbits have drastically modified much of the island's vegetation. South Georgia (54°S) has rats and reindeer, and foodstuffs imported to whaling stations resulted in the establishment of aggressive alien plants. Remote Signy Island at 60°S has experienced sporadic human presence, but the main contemporary threat is from tourism and agreement between the Antarctic Treaty managers and tour operators has exluded the island from itineraries.

The tundra has sometimes been called an

'Arctic desert' and it shares some characteristics with the biome (despite the differences in climate), including a distinct vulnerability to human activity. Both areas, too, have claims for protection as being still among the really wild places in the world.

3.3 Temperate grasslands

If we imagine a kind of temperate-zone tundra, with low vegetation dominated by grasses, and with trees only along water-courses, then we have some idea of what natural temperate grassland looks like. Usually the rainfall and snowfall (250–750 mm yr^{-1}) are too low to yield enough water to support forest but are above the level of desert, and so the grasslands are seen as an intermediate life form between the forest and the desert. The prairies, Great Plains and arid grasslands of North America, the Eurasian steppe, the veldt of Africa and the South American pampas are the main areas of this type of grassland. However, the idea of their origin as climatically determined is not always acceptable. Some grasslands occur in forest climates where perhaps a high water table favours grasses in competition with trees, or where fire prevents the

regeneration of trees and produces a grass-dominated plant community. Although lightning-set fires may play a part in producing such mosaics, it is more often the case that the fire results from human activity: early European explorers in the northeast woodlands of North America recorded large grassy openings in the forest due to fires which the Indians had started, largely to aid hunting.

The grassland vegetation can be classified on the basis of the height of the dominant species, although the grasslands are not composed entirely of the family Gramineae but a number of other herbaceous groups (notably the Leguminosae and Compositae families) are present. The moister parts of the biomes have grass species which grow more than 1 m high (in some places as high as 2·5 m), whereas the drier parts only achieve much shorter grasses. The North American grassland has been further classified into tall grass, mixed grass and short grass–bunch grass prairies, following an east–west trend of declining average precipitation and forming a gradient of falling NPP. The tall grasses (1·5–2·5 m) include the big bluestem, *Andropogon gerardi* and the switchgrass *Panicum virgatum;* the mixed grasses (0·6–1·2 m) include the little bluestem *Andropogon scoparius,* the needlegrass *Stipa spartea,*

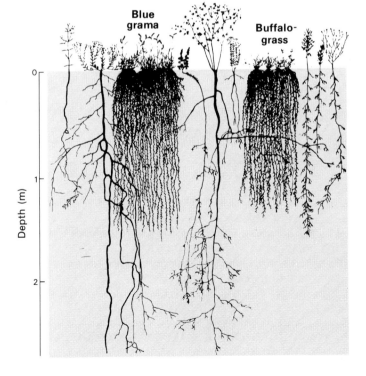

Figure 3.7 This diagram contrasts the above- and below-ground biomass of some species of prairie grasses. Dominants such as grama and buffalo-grasses exercise a strong binding effect on the soil, which is lost when the grasslands are ploughed or overgrazed to the point where such grasses disappear.

and the June grass *Koeleria cristata;* and the short and bunch grasses (0·15–0·45 m) include the buffalo grass *Buchloe dactyloides* and the blue grama, *Bouteloua gracilis.* The average biomass of these grasslands, taken together, is 1600 g m^{-2} and the average NPP is 600 g m^{-2} yr^{-1}. However, the shoots are only the obvious part of the vegetation, for the roots of most species will penetrate deeply (up to 2 m) into the soil and so the majority of the plant biomass is in fact below ground: 2000 g m^{-2} in some places (Fig. 3.7). Some of the species have underground rooting systems which help to form a very tightly knit sod which was for a long time resistant to ploughing. This sod also comprises the 50–55 per cent of the total plant biomass which annually forms part of the litter fall, so there is usually more organic matter in the litter and upper soil horizons than in the green parts of the vegetation.

The animal ecology of the grassland has some distinctive characteristics. A few species of large mammals tend to be dominant, as with the buffalo and pronghorn antelope in North America, the wild horse and saiga antelope in Eurasia, antelopes in Southern Africa and guanaco in South America. These large herbivores tend to be herd animals, which affords them some protection from predators (e.g. wolf and coyote) in the open terrain. They are usually migratory in order to avoid over-use of their forage: buffalo, for instance, appear to be rather unselective grazers (compared with cattle or sheep) although able to digest very poor quality forage. The other mammals are small, often burrowing and quite frequently nocturnal; gophers, prairie dogs, jackrabbits, and small birds such as meadowlarks and prairie chickens are eaten by weasels, foxes, badgers, owls and rattlesnakes.

All these components are linked to produce the interactive grassland ecosystem. The selective grazing of large mammals largely controls the composition of the plant community, making it more likely that an individual plant will survive if it grows steadily rather than rapidly to the point where it stands out as an obvious target for a grazer. The grassland sod holds the high quantities of nutrients and the organic matter which helps to retain moisture during long periods of drought and prevents erosion which occurs when the turf mat is broken. The

Photograph 13   The large herbivore of the North American Great Plains grassland biome, the bison. This is a protected herd in a national park in South Dakota, preserving the short-grass prairie vegetation as well as the animals.

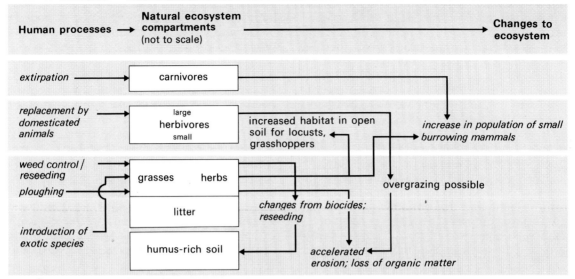

Figure 3.8   A scheme of human impacts on a temperate grassland ecosystem. The natural ecosystem is shown in the central boxes; the human activities on the left-hand side and the subsequent changes to the ecosystem on the right-hand side.

effect of man has been, in general, to break open these tight linkages, either by pastoralism to the point of grazing out the palatable species and leaving a lot of open soil whose mineral nutrients are then vulnerable to leaching, or by ploughing and exploiting the rich stores of mineral nutrients and humus (see the book on *Soil processes* by Brian Knapp in this series), sometimes with irrigation as an aid. Stable agriculture can be achieved although both the Great Plains and Khazakstan steppe, for example, have seen crop failure and soil loss due to unsuitable agricultural methods. So virgin grasslands are rare since most of them have been altered by pastoralism of domesticated animals, replaced by agricultural ecosystems, or converted to a different species composition through the use of biocides (chemical weed and/or pest killers) or mechanical processes such as brush removal, seeding with leguminous species (plants of the family Leguminosae, which have nodules on their roots containing bacteria which fix nitrogen from the air – the plants are thus independent of soil nitrogen levels), or simply through the invasion of new (including exotic) species following utilisation by man.

3.4 Tropical savannas

The tropical savanna, the first tropical biome to be considered, is in many ways intermediate between a forest and a grassland and indeed the term has been applied to a variety of vegetation formations from a nearly closed canopy woodland to a grassland with thinly scattered bushes. Common to them all is a continuous ground layer dominated by grasses. The productivity varies according to the density of the trees: the average of 900 g m^{-2} yr^{-1} conceals a range of 1500 g m^{-2} yr^{-1} in 'closed' savanna (i.e. nearly woodland) to 200 g m^{-2} yr^{-1} where the savanna is more like a desert scrub. The savanna biome is formed in a wide belt on either side of the Equator in areas with a tropical temperature regime. Total rainfall will vary from 250 mm yr^{-1} on the desert fringes of the savanna, to 1300 mm where it abuts true tropical forests, but characteristically there is at least one dry season; it used to be thought that a long dry season was the cause of the **physiognomy** of the vegetation but this feature is now recognised as the outcome of a complex of factors including geomorphic history, natural fire, the evolution of grazing animals and the presence of man. However, the dry season exerts direct effects on, for example, large herbivorous mammals, causing them to migrate in search of forage and free water. Found mainly in Africa, Australia and South America, the savanna biome covers about 20 per cent of the world's land surface.

The obvious characteristics of typical savanna vegetation are trees and grasses. The former exhibit a great taxonomic variety and are usually

6–12 m in height, strongly rooted and with flattened crowns. They exhibit drought-resisting features, including partial or total seasonal loss of leaves, water-storage modifications and reduced leaves and, in addition, are usually fire resistant (pyrophytic) in having a thick bark and thick bud-scales. In Africa, savanna trees include species of *Isoberlinia*, the baobab (*Adansonia digitata*), and the dom palm (*Hyphaene thebaica*); in Australia species of *Eucalyptus* such as *E. marginata* and *E. calophylla* form the tree layer; in South America a great variety of species is involved, and in Honduras a savanna vegetation dominated by pine trees has been described. Even where the savanna is cultivated, the trees may be retained because of their usefulness to the farmers, e.g. for shade, food, materials and firewood. The grasses are often long, reaching up to 3.5 m in height and

thus providing ample fuel for dry-season fires. In Africa, the elephant grass *Hyparrhenia* grows to 5 m in height. Other typical savanna grass genera are *Panicum, Pennisetum, Andropogon* and the African species *Imperata cylindrica*, now wide-spread throughout the Tropics because of human introduction. Only the underground parts of the grasses survive the dry season.

Where man has not over-hunted them, and where fire maintains a variety of habitats, the savannas can support a very diverse fauna. The savannas of East Africa, for example, support the greatest variety of grazing vertebrate life in the world, with over 40 species of large herbivorous mammals (such as African buffalo, wildebeeste, zebra and many antelopes) and up to 16 species grazing together, apparently in the same habitat. A wide variety of scavengers and

Photograph 14 Savanna biome in East Africa. The waterhole attracts more animals than are seen here but mostly at night. Beyond the pond is seen grassland dotted with *Acacia* trees: a habitat which is regularly burned over.

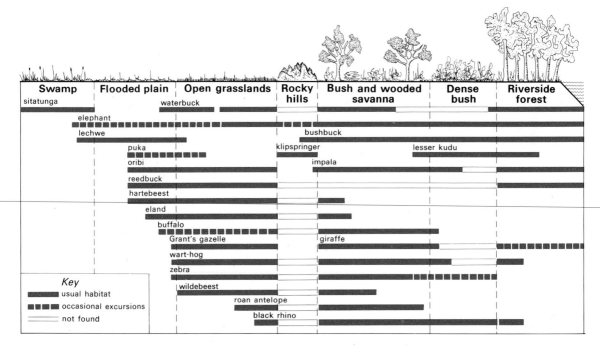

Figure 3.9   A transect of an African savanna environment, showing the variety of habitats which can be found within it. Plotted are the feeding zones of the rich herbivore fauna. Whereas many of these use the open grasslands, several have rather specialised feeding habits, which means that the environment can support them all. Even those which appear to compete may in fact partition the forage, e.g. by feeding off different heights of vegetation, or grasses at different states of dryness and seed formation.

predators is supported by this fauna. Other savannas are less rich: in many parts of Africa and South America hunting has depleted the fauna, although the known occurrence of 50 species of kangaroo in Australia suggests an evolution towards a similar faunal richness. Where a rich fauna still exists, as in East and Central Africa, it may achieve a yearlong vertebrate biomass of 100×10^5 kg ha^{-1} live weight (compare with Scottish red deer at 0.14×10^5 kg ha^{-1}) and the animals avoid over-grazing by various behavioural adaptations: they may be separated by small variations in the habitat, for example, as where one species of small antelope lives in permanent swamps, another in seasonally inundated areas. In parallel with this, some animals may be very selective in their foraging. Different beasts will use different levels of vegetation: the giraffe has sole rights to the tops of, for example, *Acacia* trees, but lower down the two species of African rhinoceros are separated by food habits, since one is a browser off twigs and foliage and the other a grazer from the

ground layer (Fig. 3.9). The interaction of the herbivores and their forage is illustrated by accurate measurements such as those from grasslands in Uganda which show that grazing mammals remove half of the above-ground NPP.

Other animals also are plentiful in savannas: birds can make good use of a variety of food species and can move quickly to evade unfavourable conditions. Year-round ground-living species and scavengers are joined by many migrant species avoiding colder winters elsewhere on the globe. The insects are dominated by the Orthoptera (locusts and grasshoppers), Hymenoptera (ants) and Isoptera (termites), and the savannas are the starting places of the outbreaks of migratory locusts which are so devastating to agriculture. Locally more important are the termites which are a major component of the decomposer chain. The importance of termites in recycling nutrients can be gauged from estimates in the Ivory Coast which suggest that the biomass of the termites, which decompose cellulose underground, is 12 kg ha^{-1} and that these consume 30 kg ha^{-1} yr^{-1} of cellulose and so yearly rearrange several dozen tonnes of surface soil, improving its structure and increasing aeration for plant growth. The mounds built by some species of savanna termite are important landscape features, with up to 600 hills ha^{-1}. They may also become the foci of trees and shrubs not palatable to the termites and such thickets are often pockets of vegetation which is not burnt, thus adding to the diversity of habitats.

The occurrence of fire in this biome has been mentioned several times. It is probably responsible for the maintenance of savannas as grassy communities, for when absent the tree cover increases and there are fewer grazing and browsing mammals. It is clearly of evolutionary significance for the ecosystems, for in Africa at least, evidence of fire in archaeological contexts goes back to 60 000 years before the present. In general, regular fire favours perennial grasses with underground stems which can regenerate after the fire has passed. The burning also actuates growth by removing dead material and mineralising litter, and the stimulation of perennial grasses in middle or late dry-season burns is particularly significant for savanna vegetation, for if the burn is early in the dry season, then the growth may be initiated but there may not be sufficient

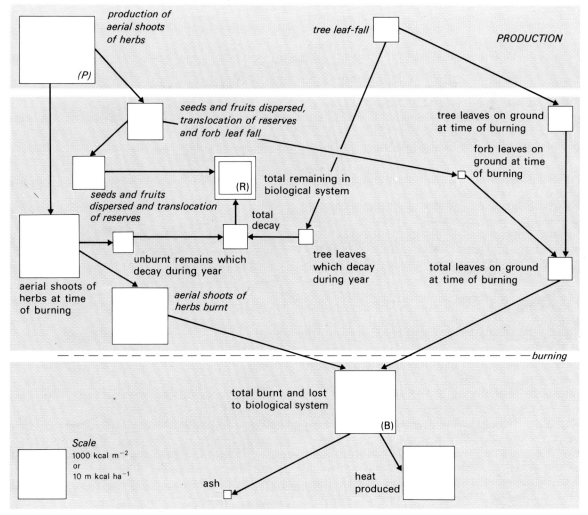

Figure 3.10   An energy-flow diagram showing the effect of burning on the savanna of Nigeria. The area of the boxes is proportional to the calorific value of the vegetation per unit area (1 cal = 4·2 J). Forbs are herbaceous plants other than grasses. The importance of burning in the system can be gauged by comparing the production box (P) with the total burnt box (B) and the total remaining in the biological system box (R).

water available to last the plant through to the rainy season. Where large mammal herbivores are absent, fire consumes what they would otherwise have eaten. Thus fire appears to be a normal part of the savanna biome and one of the major factors in its nature, as it is in many tropical forests and grasslands other than the lowland rain forests (Fig. 3.10).

Whether fire is the major determinant of the savanna biome's ecological characteristics is more controversial. There are areas of South America where savanna-type vegetation is found without fire, and where work on ecological history indicates spells of open grassland before the coming of man. In Africa, however, opinion seems to favour the idea of the savanna as a delicate balance of the outcome of climate, soils, vegetation, animals and fire, with fire as the key agent whereby men have created the biome; as it now stands this biome in Africa cannot be regarded as a climatic climax but as a product of human activity. It is less certain that this diagnosis applies elsewhere and the savanna may be an example of a similar ecosystem being the end-product of different ecological histories.

3.5 Sclerophyll ecosystems

This set of systems is named after the major adaptation of its dominant low trees and shrubs. This is the possession of thick leathery leaves with waxy cuticles and is a way of adapting to a climate in which there is a long dry season. This vegetation type, known as chaparral in California, maquis or garrigue in southern Europe and mallee scrub in Australia, is typically the vegetation of areas of 'Mediterranean' climate.

The dominants are trees and shrubs, either with needle-like leaves or with evergreen leaves of sclerophyll character. The dominants are usually 3–4 m high and form a close-set and sometimes impenetrable scrub. In the Mediterranean, the wild olive (*Olea europea*), carob (*Ceratonia siliqua*), evergreen oaks (e.g. *Quercus ilex*), pines such as *P. pinea* and *P. pinaster*, arbutus or strawberry tree (*Arbutus unedo*), heathy shrubs of the genera *Erica, Ulex* and *Genista,* and herbs of the Labiatae family and the genus *Thymus,* are the commonest. In California, many species of bush-like oak are found, as well as species of *Ceanothus,* the chamiso bush (*Adenostoma*), manzanita (*Arctostaphylos* spp.) and several pines. In Australia the equivalent mallee consists

of *Eucalyptus* scrub 2–3 m high. In an average area of the biome, a biomass of *c.* 6000 g m^{-2} results from a NPP of *c.* 700 g m^{-2} yr^{-1}: one sample site had an above-ground biomass of 315 t ha^{-1} under *Quercus ilex* and 518 t ha^{-1} of organic matter in the soil, connected by 3·8 t ha^{-1} yr^{-1} of leaf fall. It appears that many of the plants grow quite fast on soils which are low in phosphorus and so conserving mechanisms exist: a fine mat of roots penetrates the litter and enzymes within the roots accumulate the phosphorus until it is needed for rainy-season growth. The litter does not penetrate far into the soil in the form of humus and so in the soil profile the unobscured iron, coloured red, gives the name of *terra rossa* to the characteristic soils in the Mediterranean.

The abundant food and good cover of these scrub lands allow a plentiful animal life. In Southern California, 201 species of vertebrates were counted, 75 per cent of which were birds. Mammals now tend to be dominated by ground squirrels, the wood rat and the mule deer (*Odocoileus hemionus*) although before the heavy impact of man, predator species such as the wolf and mountain lion, and **diversivores**

Photograph 15 Schlerophyll vegetation (chapparal, maquis, garrigue, are all common terms) **in Sardinia.** All the dominants are of low height and all are adapted to a long dry season. Fire is a common factor in such ecosystems: its frequency has increased in recent years as human impact upon such systems has increased.

(animals able to eat a wide variety of foodstuffs) such as the grizzly bear (*Ursus arctos horribilis*) were more common. The role of small mammals in the ecosystem is hinted at by the fact that wood rats (*Neotoma fuscipes*) have been known to consume the entire acorn crop of one species of oak in a particular year.

In such climates, it is not surprising that fire is a normal occurrence in the biome. In the San Dimas forest of California, lightning set eight fires in 75 years and burning by the Indians is well documented. Most of the species are adapted to fire (e.g. after it, eucalypts give off numerous stems from the stump, like a coppiced tree) and no doubt have been selected by many thousands of years of its occurrence. The fires seem to stimulate the germination of some seeds, to reduce to ashes much vegetation and litter and hence speed up the process of mineralisation of organic matter; and also to destroy **phytotoxic** compounds (poisonous to other plants and perhaps also to bacteria and other soil organisms) secreted by plant roots which interfere with litter decomposition and the processes of nitrogen fixation in the soil. Its role in succession is illustrated by an Australian study which showed that an unburned *Eucalyptus* forest was experiencing no regeneration of the Eucalypts but that they were being replaced by *Casuarina, Banksia* and *Acacia* species.

Chaparral and similar forms of vegetation clearly can be regarded as biomes in which natural fire has played a key role, even though its frequency has been increased by human activity. Fire, though, is not the only form of human impact; centuries of grazing of goats and sheep in the Mediterranean, terracing for cultivation, management for high deer populations, conversion to grassland using biocides, agricultural use including irrigation, coppicing and urbanisation have all hit the sclerophyll scrubs at various times, some of them quite recent, e.g. homebuilding in the Santa Maria mountains near Los Angeles, others quite ancient such as pastoralism in the Levant since time out of mind.

3.6 Boreal coniferous forests

With this formation, we come to the first biome dominated by trees. In spite of long and cold winters with considerable snowfall and soils frozen to a depth of 2 m, the evergreen conifers thrive. They are helped by a summer which,

although of short duration, has a long day length and at least one month with an average temperature of 10°C.

In such a physical environment, the evergreen conifer has particular adaptations suited for survival: provided water is available it can photosynthesise all year round, and the needle-leaves help to resist drought when water is locked up as ice and when strong winds increase transpiration rates; the shape of the crown enables it to shed snow so that branches are not broken off by the weight. Large trees up to 40 m high are common and these dominate the structure of the biome: in a continuous stand relatively little light penetrates the canopy so that a lower layer of trees is uncommon (Fig. 3.11). There is usually a continuous ground cover, the components of which vary according to local conditions of drainage and light: at the dry end of the spectrum a lower cover of lichens and mosses may be found; where there is more ground-water the vegetation will include low heath shrubs such as crowberry; and where it is very wet bog-mosses such as *Sphagnum* will cover the ground, often forming open bogs in very wet hollows.

The great dominant genus of the biome is

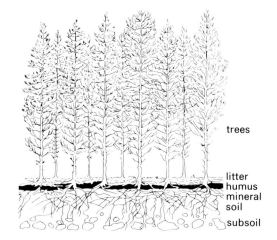

Figure 3.11 A sketch through a boreal coniferous forest. There is only one tree layer, the shading effect of which is to preclude the growth of any lower tree layers. The litter breaks down slowly and produces a deep litter layer; slow mixing of the humus with the mineral soil gives a sharp break between humus and soil. Food for larger animals is sparse in such an ecosystem and they are more likely to be found where fire or windthrow has created gaps in the forest which are being colonised by deciduous shrubs such as birch, aspen and willow.

Photograph 16 Boreal coniferous forest. The forest is dominated by a single species (in this case a pine) and there is a heathy ground layer. Such ground layers, coupled with an acid humus derived from the needle-leaf litter, often make regeneration of seedlings difficult and fire sometimes helps in their establishment by thinning the humus layer and mineralising some of the nutrients in it.

Picea, the spruces. The forest tends towards single-species stands and so great swathes of spruce are found, with more diversity near breaks in the forest such as streamsides, areas of windthrow and burned areas. The firs (*Abies*), pines (*Pinus*) and larches (*Larix*) tend to be more restricted in their occurrence in the main sweep of the biome, though the larches, for example, are very important in Siberia. Deciduous species, such as birch, willow and aspen, are also found in this biome. One of their localities is at the northern or altitudinal limit of the coniferous forest: in Lapland, for example, a broad belt of birch scrub separates the evergreens from the tundra-type vegetation, and in the Rocky Mountains the aspen (*Populus tremuloides*) occurs right at the tree-line before the alpine grass and herb communities take over. The second role of the deciduous species is as pioneers in deforested areas – as when windthrow has opened up the coniferous blanket, or when fire has removed the dominants, or where succession is kept at an early stage along rivers subject to flooding and the shifting of sand and shingle banks.

The size of the dominant trees and their ability to photosynthesise all year round if conditions permit, mean that the NPP of this forest is sometimes not far short of forests further to the south. In northern Japan, for example, the NPP of firs–spruce forests averaged 2000 g m^{-2} yr^{-1}, whereas deciduous forests to the south averaged 2160 g m^{-2} yr^{-1}. But the average for the whole biome is much lower at 800 g m^{-2} yr^{-1} (that for deciduous forests is 2000 g m^{-2} yr^{-1}) so that these forests must have been unusually rapid in their growth. The seasonality of the climate is strongly reflected in the breakdown of the litter, which is slow compared with other biomes and often results in a deep accumulation of organic matter on the forest floor, especially in wetter areas. Up to ten times the annual litter fall may accumulate on the forest floor, so that a litter biomass of 100–500 kg ha^{-1} is found. (In deciduous forests the equivalent measurements are five times the annual fall and 100–150 kg ha^{-1}.) The slow mineralisation of this material is short-circuited by fire which under natural conditions is probably quite frequent and runs along the forest floor consuming the relatively low amounts of litter accumulated since the last fire. Man-introduced fire-protection policies allow debris to accumulate and so a big fire may then result, with the fire running up the trunks (aided by resinous drips) and igniting the crowns. Such a fire is disastrous in commercial terms but probably infrequent under natural conditions. A shallow burn has another role of reducing humus thickness so that seedling trees can more easily root in the mineral soil, and to that extent it may aid forest regeneration. The invertebrate animals of the soil have a considerable capacity to affect the soil conditions and the regeneration of trees, for example, and so probably are more important members of the ecosystem than the more obvious animals. The biomass of the coniferous trees yields only a small amount of available food so that in a forest of some 200 kg ha^{-1} of tree biomass, the above-ground animal biomass of vertebrates might be only 2·5 kg ha^{-1}.

Of the vertebrates, rodents are a characteristic group and generally survive the winter under the insulating cover of the snow blanket. The beaver (*Castor fiber*) is typical of this biome and is perhaps analogous to the elephant in its alteration of the local habitat. Of the larger mammals, various sorts of deer are characteristic including the moose (*Alces alces*) which is largely an animal of the secondary vegetation produced by fire. The diversivores are typified by bears, including the common brown bear (*Ursus arctos arctos*) as well as the remnant populations of the North American grizzly (*Ursus arctos horribilis*). Small carnivores such as lynx (*Felis lynx*) and wolverine (*Gulo gulo*) prey chiefly on the rodents, as do the owls and hawks; the characteristic large carnivore is the wolf (*Canis lupus*) which is an important predator upon the populations of deer, caribou and moose.

Man has used this biome ever since it was formed. It was the basis for many hunting and gathering cultures based on deer, moose and fish. Now it is the focus of the world's softwood lumber industries whose activities alter the ecology considerably – sometimes temporarily and sometimes over a long period, especially when the frequency of fire is increased beyond that to which the forests are undoubtedly adapted. Also, much of the nutrient supply of the ecosystems is tied up in the biomass and so rapid-cycle exploitation and whole-tree harvesting may reduce the productivity of this biome permanently.

3.7 Temperate deciduous forests

To the south of the boreal forest is found a more productive forest type, but one in which there is not growth all the year round; the trees loose their leaves (the **deciduous** habit) in the winter, so that there is a dormancy period as far as the dominants are concerned. These forests (found in western and central Europe, the northeastern USA and Japan south of Hokkaido) respond to a climate having marked hot and cold seasons with winter temperatures which fall below the freezing point of water; the precipitation is of the order of 750–1500 mm yr^{-1}. The deciduous habit of the dominant trees can be seen as a form of **dormancy** which is a seasonal response to low energy levels from the Sun and the winter freezing of water. In spite of this, NPP averages at 1200 g m^{-2} yr^{-1}, which is not much less than evergreen forests in the temperate and boreal zones. The distribution of this biome is confined notably to the northern hemisphere: areas of similar climate south of the Equator are occupied by evergreen forest.

The biome is dominated by trees of 40–50 m height. Their leaves tend to be broad and thin (compared with the leathery but narrow

Photograph 17 An oak–beech wood of the deciduous forest biome, in France. Below the dominant trees, there is enough light for some shrubs and regenerating saplings of the dominants. There is sufficient light, too, for a ground vegetation of herbaceous plants to develop.

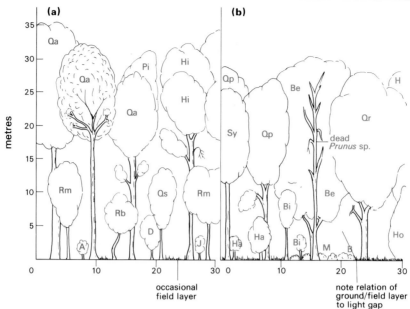

Figure 3.12   Profiles of deciduous woodlands in (a) Virginia, USA, and (b) Hertfordshire, England. The letters refer to individual trees of different species. Note the layering of trees, where the canopy allows an understorey layer of trees (mostly of different species from the main canopy-formers) to grow. The field or ground layer vegetation is closely related to where gaps in the canopy let in additional light. Both transects are 10 m wide.

leaves of tropical genera) and a large number of them produce nuts and winged seeds rather than pulpy fruits. Acorns, beech mast, chestnuts and the winged seed of the sycamore are examples of this habit. Dispersal by animals becomes very important if the seeds are too heavy for transport by wind, and indeed the density of herbivores (e.g. mice and voles) may affect strongly the regenerative success of the trees. These trees are occasionally the habitats of climbers such as ivy (*Hedera helix*) and wild vine (*Vitis* sp), and **epiphytes** such as mosses, lichens and algae grow on the trunks. Dominance by two to three species is common and single-species stands are often found. A dense canopy is usually formed and the amount of light percolating through determines the character of the lower layers of vegetation, as does the leaf-mosaic density of the individual species of dominant tree (Fig. 3.12). The dominant trees are typically oaks (*Quercus* spp.), beeches (*Fagus* spp.), elms (*Ulmus* spp.), limes, lindens or basswoods (*Tilia* spp.), tulip trees (*Liriodendron* spp.), chestnuts (*Castanea* spp.), maples (*Acer* spp.) and hickories (*Carya* spp.), though the species variety per unit area never approaches that of the Tropics. In North America, the maximum is about 40 spp ha^{-1}, in Europe 8 spp ha^{-1}. Soil parent

material may affect the distribution of forest-tree species; in Britain freely drained gravels are associated with woodlands of *Quercus petraea* (oak), *Fagus sylvatica* (beech) and *Carpinus betulus* (hornbeam), whereas loamy soils, derived for example from clays, favour another oak (*Quercus robur*), elms and lime. Thin soils over chalk are also associated with beech and hard limestones often carry woodland fragments of ash (*Fraxinus excelsior*). Wet soils are tolerated by willows (*Salix* spp.) and alder (*Alnus glutinosa*).

Below the trees a shrub layer may form, especially where light penetrates in gaps in the forest canopy, and a great variety of species may be involved depending on the continent. Birches (*Betula* spp.), ash (*Fraxinus* spp.), hazel (*Corylus* spp.), members of the Rosaceae (e.g. the genera *Prunus, Rosa, Rubus*), grow in the gaps and form part of the succession back to mature forest. Some of these shrubs are spiny or have prickly leaves (e.g. holly, *Ilex aquifolium*) and may thus protect the seedlings of forest trees from browsing mammals. Often the seasonality of the forest is reflected in the ground flora which exhibits two assemblages: an early spring group which puts out leaves and flowers and which sets seed before the dominant trees have come into leaf; and a summer group which can tolerate the lower light levels of the canopy in full leaf. In Europe, the wood anemone (*Anemone nemorosa*) is an example of the former and dog's mercury (*Mercurialis perennis*) of the latter. Bracken fern (*Pteridium aquilinum*) is also characteristic of gaps in the canopy. A lower ground layer of mosses and, on very dry sites, lichens may also be found, and the presence of this stratum is less obviously dependent upon the summer light intensity.

The animal communities too are responsive to the climatic regime. Migration to warmer climates for the winter is common among insect-eating birds such as the warblers; hibernation is found among those less capable of long-distance travel, such as the black bear. Others, e.g. the deer of various species which are chiefly browsers, remain active all year, digging through the snow for food when the above-snow browse is exhausted.

The litter layer is the site of a diverse flora and fauna which are responsible for the release of inorganic mineral nutrients back to the soil. There are two main stages in this process. Firstly, the primary decomposers (millipedes, woodlice, beetles and earthworms) attack the litter and break it down into smaller particles. In the second phase, these fragments, together with the faeces of these litter animals, form the food for the secondary decomposers (mites and springtails) which further comminute organic material. Wet material is broken down by bacteria, fungi and protozoa and so more or less complete mineralisation is achieved: there is little long-term accumulation on the surface, although some centimetres of humic material are present nearly always.

The energy and nutrient flows of deciduous forests have been studied more intensively than for many other biomes. We know, for example, that the accumulation of biomass in North American forests seems to level off at *c.* 400 t ha^{-1} when the forest is 200 years old; European forests of comparable age seem to have less biomass. Mineral cycling has also been studied intensively. The minerals in the shed leaves are clearly important in the supply of nutrients to soil fauna and flora and hence to the trees again. As might be expected, though, the rate of uptake of the nutrients into the trees is much lower than in tropical forests – at 200–500 kg ha^{-1} yr^{-1} about 25 per cent of the value of tropical rain forest.

Hence the soil is a proportionately much larger repository of minerals than in tropical forests. Losses from leaching are therefore high by tropical standards but the nutrient cycle can still be described as 'tight'. In temperate climates, losses to runoff are made good from inputs in precipitation and by the relatively fast weathering of the rocks, and the minerals are lifted into the biosphere by trees with deep taproots.

No account of this biome can over-emphasise the effect of man, for it is in this zone that Western industrial civilisation grew up, preceded by a long period of agriculture. There have been two main effects: the first of these is obviously the clearance of forest and its replacement by agriculture. Notably, the resulting ecosystem had a greater proportion of its nutrients in the soil (as distinct from the vegetation) than in the Tropics, so that fertility remained high, particularly where the annual leaf fall was mimicked by annual inputs of animal manure from mixed farming; the dung helped to maintain the crumb structure as

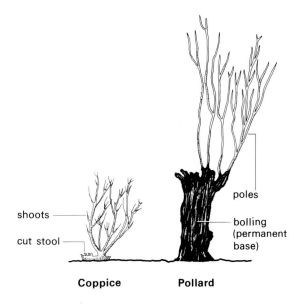

Coppice   Pollard

Figure 3.13   Two techniques of tree management once prevalent in Europe. On the left a tree is cut down to its base (stool) and then a dense mass of sprouts produces a crop of poles for agricultural and building purposes in a few years. This is called coppicing and hazel was often so treated. On the right a major tree is cut above the reach of browsing animals and again a dense crop of useful poles is produced. This is pollarding and was applied to many forest trees, for example oak, beech and hornbeam.

well as the nutrient levels (see *Soil processes*). Secondly, the remaining forests have been heavily managed for an immense variety of purposes, from timber production through forage for domesticated herbivores to amenity in terms of visual pleasure or hunting. The effects on forest ecosystems are still being unravelled by historical ecologists but it is clear, for example, that the demand by shipbuilders for timber in Europe in the 16th to 18th centuries encouraged a move towards single-species woodlands of oak, just as demand by furniture-makers in the 18th and 19th centuries encouraged single-species forests of beech. Earlier still, common rights legislation enabling villagers to take wood 'by hook or by crook' may have brought about pollarding; and a rural society's demands for pliable but straight poles produced many an area of coppice (Fig. 3.13). Species such as hazel (*Corylus avellana*) were often used because of the concomitant nut yield, but oak coppice and lime coppice were also found. The upshot is that it is unlikely that any deciduous woodland in Europe is 'natural' (i.e. unmanipulated by man), although it may be 'primary' in the sense that woodland has grown on that site from time immemorial (i.e. at least since the Dark Ages). So if a forest is to be called 'natural' it needs to have a certificate of its pedigree which shows that for both 'prehistory' and 'historical times' all the techniques of the ecological historian have failed to reveal any trace of human effect.

In Britain, it seems certain that many of our areas of 'wild' vegetation have been differentiated out of a forest which covered the country in prehistoric times (Fig. 3.14). So many open vegetation communities which look natural in the sense that they are covered with wild and not cultivated plants are in effect man-made communities. We know, for example, that during prehistoric time, human groups cleared forest from most of the uplands of the country and that their descendants, using fire and grazing animals, kept the areas free of trees. This applies to areas such as Dartmoor, the Pennines, much of Wales and highland Scotland. Lower areas too have had much the same history. Heathlands were once forest and so was much of the downland associated with the chalk in southern England, both areas suffered clearance during prehistoric times and then were kept treeless by various combinations of grazing and burning (heathlands) and grazing and cultivation on the chalklands. One result is that in Britain, apart from perhaps a few salt marshes and some remote sand dunes, we have no really natural ecosystems: even the vegetation of the high plateau of the Cairngorms, for instance, probably has been changed by different densities of red deer influenced by hard management policies.

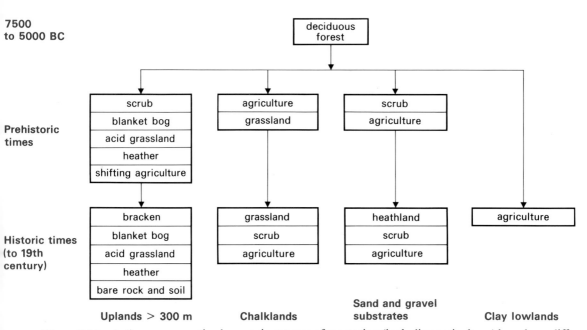

Figure 3.14 A diagram suggesting how various types of vegetation (including agriculture) have been differentiated out of an original (7500–5000 BC) cover of deciduous forest in England and Wales. The 20th century has added land uses which are not treated here and which are often independent of substrate or altitude.

3.8 Tropical evergreen forests

Conditions most favourable to high productivity seem to have developed in lowland equatorial zones. The constant high temperatures and year-round precipitation have allowed the development of large trees with rapid growth rates, high speed litter breakdown and fast nutrient cycling.

The tropical rain forests are found in the lowland areas of the Zaïre and Amazon basins and in archipelagic and peninsular southeast Asia with extensions into lowland Meso-America and Madagascar. They occur in climates which have both high and constant temperature and humidity, with precipitation of over 2000 mm yr^{-1} and at least 120 mm in the driest months. The vegetation is dominated by trees of a great variety of species – in some parts of Brazil there are 300 different species of tree in 2 km^2. The trees typically are tall and are structured into three layers (Fig. 3.15). The highest or emergent layer consists of the tallest trees (45–50 m high) which are scattered but project through the lower canopy layer (25–35 m high) which forms an almost continuous cover and which absorbs some 70–80 per cent of the incident light. When there are gaps in this lower layer, the normally sparse understorey tree layer may become dense. The

effect of these tree strata is to absorb all but a few per cent of the incoming light and so shrubs and ground vegetation are not normally found except in gaps in the forest and at its edges and near rivers. The trees themselves typically have buttressed bases for mechanical support, are evergreen with leathery green leaves and are **aseasonal** in that they both flower and lose leaves all year round. They are rarely wind pollinated because of the species diversity, but rely instead on a variety of insects, bats, birds and even mammals: in Madagascar, lemurs appear to be important in pollinating various plants, including trees, and the same seems to be true of the bushbabies of continental Africa.

Examples of the tree species in this biome are the rubber tree (*Hevea brasiliensis*), wild banana (*Musa* spp.) and cocoa (*Theobroma cacao*). Dominance may be difficult to detect since family dominance may replace species dominance in areas of high species diversity: in the lowlands of Malaya the family Dipterocarpaceae provides perhaps 25 per cent of the mature individuals, and in Borneo the large Dipterocarps may constitute 90 per cent of the emergent layer. However, single-species dominance is found as well: in South America genera like *Mora, Eperua* and *Dimorphandra* form such stands, and their

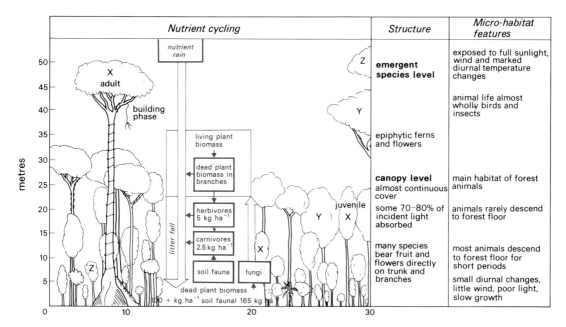

Figure 3.15   A profile diagram through tropical rain forest showing some aspects of both its structure and function. Some of the lower levels of trees are immature specimens of emergent species and they are marked X, Y and Z. Note that the animal life is also stratified but that its biomass is small compared with the biomass of the trees. A great deal of the biomass is tree trunk and branches and this provides little food for animals.

Photograph 18 A rain forest tree in Tikal (Guatemala), with the buttress roots typical in this biome. In this case, there are some shrub-height trees but notice the absence of ground layer and its replacement by abundant litter.

equivalents are found in southeast Asia and Africa. This type of dominance (common in temperate zones) was thought to be a feature of poor soils, but more recent work has put forward the idea that very high tree diversity in rain forests is a feature not of the mature forest but of a successional state due to past natural or man-made disturbances and that in an equilibrium or mature state the tree diversity of the forest is lower.

Animal life, though very low in productivity compared with the trees, exhibits the greatest taxonomic variety of any biome. The richness of food resources available and the relative constancy of environmental conditions seem conducive to such a state. Like the trees, the animal communities are stratified: the emergent layer is inhabited mostly by birds and insects which live their whole lives in this arboreal habitat. Below them, the canopy layer houses the highest variety of animals in the forms of tree-dwelling monkeys, sloths, anteaters and small

carnivores. They rarely descend to the ground, but in the understorey layer the animals may range down from the trees to the forest floor. Ground-dwellers are less diverse than arboreal types but include deer, rodents, peccaries and wild pigs. The species diversity is most obvious perhaps among birds, where the abundance of fruits, seeds, buds, nectar and insects allows dense populations of groups such as hornbills and toucans, and the presence of birds-of-paradise.

The productivity of the rain forests seems to depend upon mechanisms which keep the nutrients in the organic components of the cycle so that they are not leached out of the inorganic phase by the abundant rainfall. The rate of litter fall from the forest canopy is high (typically 11 t ha^{-1} yr^{-1} in the Amazon basin) but there is a humus turnover of 1 per cent per day so that litter does not accumulate: if mean temperatures are above 30°C, litter is broken down faster than it is supplied, at 25–30°C supply and breakdown are about equal. The main agents of litter breakdown seem to be fungi in **mycorrhizal** associations with the tree roots, so that mineral nutrients are passed directly from the decaying litter to the roots of the trees for uptake. Thus loss of minerals to the runoff is minimised, even to the point where soil animals are forced to feed on the fungi rather than the litter. Earthworms, for example, do little mixing of the soil and so the organic upper horizon is sharply marked off from the mineral soil beneath. However, the energy-rich litter does sustain a dense population of bacteria and blue-green algae which are nitrogen fixers and these are essential maintainers of a nitrogen cycle that plays a key role in sustaining the high biomass, though the circulation of large quantities of silicon, calcium and potassium is important as well. Input from precipitation is by no means negligible for nutrients either: near Manaus, Brazil, the annual average input from rainfall was 0·3 kg ha^{-1} of phosphorus, 2·0 kg ha^{-1} of iron, 10 kg ha^{-1} of nitrogen and 3·6 kg ha^{-1} of calcium.

The outcome of high solar input, abundant rainfall and rapid nutrient cycling is a very high NPP, with the mean for rain forests estimated at 2200 g m^{-2} yr^{-1} of dry matter; multiplied by the area of the biome, we get 37·4 × 10^9 t yr^{-1} which is far higher than for any other terrestrial biome. The average subsumes the fact that there are large areas of rain forest on podsol soils which fall at the lower end of the productivity range (1000 g m^{-2} yr^{-1}) for this forest type.

Nutrient-cycling studies show that the clearance of this type of forest for agriculture robs the system of its store of nutrients and hence its fertility in agricultural terms. Once the trees are gone, nutrients are lost in runoff after being rapidly mobilised in the higher temperatures of the now unshaded soil surface. It is notable that traditional shifting cultivation adapted to these conditions by planting a miniature forest of crops which covered the clearing surface – tidy row crops were avoided. The plot was abandoned after a few years to enable the forest to recolonise and rebuild the nutrient cycles. Such a strategy is not, however, the aim of large-scale clearance schemes which involve keeping the forest down, once felled, by grazing domesticated stock. Various governments and lumber companies are removing the forests at a current rate of 11 million ha yr^{-1} so that of an original area (about 5000 years ago) of 16 million km^2 of rain forests, only two-fifths are now left: at this rate of destruction, another 30 years will see the demise of the biome. Apart from its scientific interest (for example there are estimated to be over 25 000 species of flowering plants in the rain forests of southeast Asia), such a biotic profusion must inevitably be a reservoir of great economic and genetic potential.

Beyond the apparently ideal conditions of lowland equatorial regions (e.g. where rainfall becomes more distinctly seasonal), other, modified, types of forest are found within the Tropics. In general, they lack the triple layering of the lowland rain forests and are more likely to be deciduous, or at least mixed evergreen–deciduous, depending on the moisture-retaining capacities of their soils. Single-species stands are more common, and include those of the most famous tree of these biomes – the teak (*Tectonia grandis*). There is more frequently a shrub layer than in true rain forest and this, coupled with a dry season, may mean the accumulation of litter on the forest floor, a feature uncommon in the lowland equatorial forests. Dryness plus litter provide good conditions for fire and so this factor enters the ecology of these forests, preventing their colonisation, for example, by some of the species of the evergreen rain forest, which must be one of the very few terrestrial biomes in which fire seems to play an insignificant role. Nevertheless, the mean productivity of seasonal tropical forests is 1600 g m^{-2} yr^{-1}, which places them second only to the equatorial lowland forests in the ranking of terrestrial NPP.

3.9 Islands

It seems appropriate to bridge the descriptions of terrestrial and aquatic biomes with a discussion of islands. Here we must remember that the sea is likely to be a strong biological influence, even on apparently terrestrial fauna and flora. In the Arctic, for example, many small islands bear numbers of animals that could not possibly be nourished from their sparse tundra, but which depend on the sea for their nutrition, either directly or indirectly. Islands tend to exhibit certain defineable biological characteristics: they have a low species diversity compared with the nearest continental masses, a diversity which gets progressively lower away from the continents, especially along island chains; they show adaptive radiation, where a restricted taxonomic group evolves to fill a large number of niches, as with the finches of the Galapagos, or the lobelias of the Hawaiian islands; there is constant immigration and extinction as the suite of plants and animals proceeds towards an equilibrium; and they exhibit a tendency for early stages of succession to be unusually common compared with the mainland because of the incidence of catastrophes which may affect all or much of the island, e.g. volcanic eruptions, hurricanes and, to a lesser extent, tidal waves.

Island ecosystems will, of course, vary with the usual environmental factors, and the relief and substrate are major determinants since the term island encompasses the sort of variation that exists between islands such as Hawaii, with the active volcano of Mauna Loa at 4170 m, and low atolls never more than 15 m above sea level. In between there are many geologically more 'normal' islands with a great variety of terrain. In all of them the coastal ecosystems (which in the case of small coralline islands may be their entirety) are intimately linked with the sea, which may provide nutrients, for example via bird guano.

The island ecosystems are not necessarily unique in terms of their structure. For example, the island of Hawaii (3200 km from the nearest continent and 720 km from the next island group) possesses natural formations of evergreen rain forest, evergreen seasonal forest, savanna, grassland, scrub, alpine tundra and near-desert. These are arrayed in a general altitudinal sequence similar to continental tropical mountains, and the plants often exhibit similar forms and in some cases similar species. The moss *Rhacomitrium lanuginosum* occurs in the alpine tundra on Hawaii as it does on the Cairngorm Mountains of Scotland, for example, and the tree-like *Senecio* of continental tropical mountains is paralleled in the same zone on Hawaii by the silverswords (*Argyroxiphium* spp.).

The island communities are likely to be different at the level of fine structure: the species assemblages will differ markedly from their continental equivalents and there may be different spatial relationships along environmental gradients because of the poorer diversity of fauna and flora available to make up the ecosystems. For example, on the island of Hawaii there is only one woody vine in the evergreen forests, together with a species of shrub which sometimes will grow as a vine but only up to 5–8 m. Similarly, the pioneer stage of colonisation of new volcanic surfaces in the rain-forest zone includes no native grasses, their place apparently being taken by ferns.

Usually the effect of man upon island ecosystems has been strong. This is because animals and plants have been brought in which are competitively superior to the native species, or because the paucity of the island's species diversity meant that there were unoccupied places in the ecosystems. Also, since islands often have many endemics (i.e. species found only on one island or island group) in their flora and fauna, it was often easy to bring about extinction – the example of the dodo is perhaps the best known.

Animals introduced to islands are dominated by sheep, goats, pigs and rats, often left to breed on an island by sailors who fancied fresh meat on the way home. Rats, for example, may kill easily the flightless land birds which are common elements of island faunas; if they render them extinct then they can turn to other species, moving if necessary to eat chicks or eggs rather than adult birds. Grazing animals will eat out palatable plants, sometimes leaving only a semi-arid scrub which may contain introduced plants which are resistant to grazing. Such introduced fauna are a problem on islands of world interest such as the Galapagos – which has a fauna displaying many of the features which led Darwin to his ideas on evolution – and they occupy a critical place in the history of ideas. Islands often suffer, too, from being convenient for military bases since they are staging points in long stretches of ocean; some years ago Aldabra,

which has, among other animals, endemic giant tortoises, was threatened in this way.

3.10 The seas

As an environment for life, the world's salt-water bodies, which constitute 70 per cent of its surface, are very different from the land surfaces which they surround. The ocean basins are deeper than the land is high and life extends to all depths, although the biota are much more concentrated at the interfaces with the land. The oceans are all connected, forming one continuous water body so that its temperature, depth and salinity form the main barriers to the movement of marine organisms. Another difference is that all terrestrial organisms have to cope with the strain of gravity, whereas biota suspended in sea water are freed from the brunt of a requirement of energy and nutrient investment in bony or woody tissues.

The outcome of both the opportunities for and constraints upon life in the seas is a highly variable NPP. In essence, the near-shore environments such as estuaries and coral–algal reefs are highly productive (since in part the movement of water brings in food and removes wastes so that the organisms spend less energy acquiring food and getting rid of wastes), the continental shelf areas (especially where water wells up from the ocean floor bringing a nutrient supply) next in productivity, and the open oceans rather low (Table 3.1).

These figures suggest that one of the main factors limiting growth in the seas is the supply of nutrients: where these are plentiful and in the surface zone of water into which light can penetrate (the **euphotic** zone, < 200 m deep, and there are no terrestrial ecosystems with such a deep photosynthetic zone) then productivity can be high. Elsewhere, meaning for most of their extent, the seas resemble a very wet tundra (Fig. 3.16).

Table 3.1 NPP of marine ecosystems.

	Mean NPP ($g\ m^{-2}\ yr^{-1}$)	Nearest terrestrial equivalent
estuaries	1500	tropical seasonal forest
coral-algal reefs	2500	tropical rain forest
continental shelf	360	c. 60 per cent of temperate grassland
upwelling zones	500	temperate grassland
open ocean	125	tundra

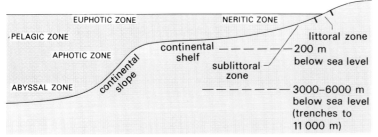

Figure 3.16 The life zones of the sea. All the photosynthesis takes place in the euphotic zone and the aphotic zone is permanently dark. The majority of life in the seas is found between the shores and the edge of the continental shelf; beyond this limit the productivity of the waters is akin to that of deserts.

The primary producers of the oceans are predominantly phytoplankton down to a depth of 60 m, although large algal seaweeds may have a very high NPP, especially in middle latitudes. Often the phytoplankton are divided into net plankton (which can be caught in the finest silk net of aperture size 0·06 mm) and nanoplankton (which pass through the smallest possible mesh). The nanoplankton, especially tiny green crea-

tures with swimming organs like tails from 2–25 μm (1μm = 10^{-3} mm) in size, seem to be the most important photosynthetic organisms of the oceans, and as such form the basis for the food chains (Fig. 3.17). The first consumer level is that of zooplankton, which feed either directly on phytoplankton or on detritus derived from them. Then there are the carnivorous zooplankton, and thereafter the many trophic levels of the long food

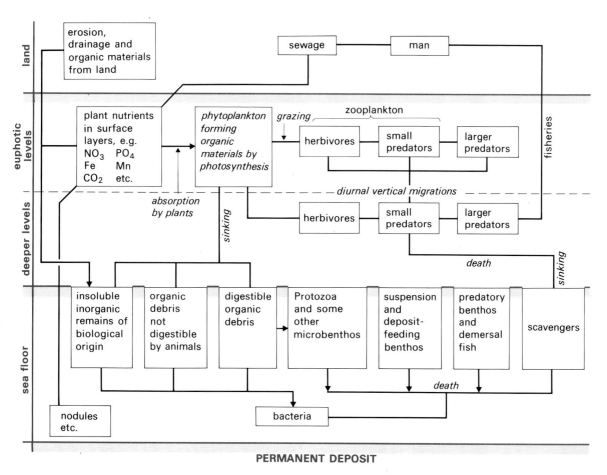

Figure 3.17 A scheme of a food web in the seas, arranged vertically with the surface of the sea near the top. A great many stages in the open water are missed out but enough remains to suggest that food webs of some complexity and length must exist.

Photograph 19 Marine plankton, plant and animal, at a magnification of × 170.

are thought to play a similar role in nutrient release to that in the litter layer of terrestrial ecosystems. The nature of the substrate exerts a strong effect on the actual species found on the sea floor of the continental shelves, but on each type of substrate the secondary productivity declines with depth, although the species diversity often increases even down to the very deep waters.

The active swimmers of the shelf zones comprise fish, the larger Crustacea, turtles, marine birds and mammals. Although individuals may range over a wide area, they are still limited in their distribution by the barriers of temperature, salinity and nutrients that affect the other components of their ecosystems. Even if not directly affected, they are tied down to their food sources. Because of the small size of plankton, fish which eat them are important links in the food chains over the continental shelves: the herring family (herring, menhaden, sardine, pilchard and anchovy) are very important in this and in the Pacific some sardines are virtually herbivores. As the fish get larger their food sources may change to smaller fish, so that long food chains with tertiary and quaternary carnivores may be found. Most marine fish lay large numbers of eggs which receive no parental care and indeed are added to the food supply of the free-swimming organisms. These fish also show a tendency to aggregate, or form 'schools', and to make seasonal migrations. This latter habit, with selective feeding even of zooplankton, probably prevents 'over-grazing' of plankton as well as ensuring adequate nutrition of the fish. Birds are also important high-level predators of these seas; they spend some time on shore and thus return some of the nitrogen and phosphorus to the land from which it came – completing a nutrient cycle.

The open oceans beyond the continental shelves are populated entirely with open-water and sea-floor organisms. The oceanic phytoplankton are predominantly very small and the zooplankton are the permanent kind without the addition of larvae of other groups. Life in open waters is aided by flotation mechanisms such as spines, fat droplets, gelatinous capsules and air bladders, and the general paucity of nutrients is fought with recycling mechanisms which keep nitrogen and phosphorus, for example, closely bound into organisms rather than loose in the open water. Even though plankton productivity is low, it eventually supports a characteristic fauna of oceanic birds such as petrels, albatrosses, frigate

chains of the seas, including fish, comprise carnivorous animals; there are few large animals which are strictly herbivorous. The Crustacea constitute about 70 per cent of the zooplankton (if the temporary contribution to the zooplankton of the larval phases of many sea animals is excluded) and include the important group of the copepods, about 2000 species of mostly herbivorous zooplankton. Their role gains significance from the estimate that at least half the NPP of the oceans is converted for a time into wax and that the copepods are the main producers of wax in the oceanic ecosystem.

The larger consumers of the sea bottom are sometimes free moving (as with some flatfish and lobsters), or may be fixed (sea-anemones, bivalve molluscs), or may burrow into the substrate if it is sand or mud (burrowing anemones, bivalve molluscs, gastropods, echinoderms and Crustacea). Bacteria are also found in large numbers in the surface sediments of the sea floor where they birds and terns which only come to land in order to breed; most of these are bound to a particular type of surface water even though their movements appear to be unconstrained. Completely independent of the land as well are sea mammals such as dolphins and whales. Zooplankton are the main food of most large whales, and free-swimming animals such as squid are the major prey of the toothed whales.

Compared with many terrestrial ecosystems, the dynamics of marine populations are so little known that the results of human interference are often hard to elucidate, apart from spectacular examples such as the decline of whales from over-hunting (Fig. 3.18). The effect on the seas of nutrient enrichment from sewage and fertilizer runoff, for instance, is complex and affects both species composition and productivity in a number of ways; analogous are the fate and effects of long-lived substances such as chlorinated hydrocarbon biocides (e.g. DDT) and polychlorinated biphenyls (PCBs) although the toxic effect of these upon carnivores such as sea birds is now well documented. Contamination of the sea has resulted in the excessive growth of **sessile** green algae which in turn has underlain the depletions of the oxygen content of the bottom of shallow seas such as the Baltic, and so a permanent zone without oxygen – the bacteria of which produce large quantitites of hydrogen sulphide (H_2S) – has become established. The hydrogen sulphide causes the release of phosphorus which adds to the

Photograph 20 An algal–coral reef fringing the island of Viti Levu in Fiji. In many places the fringe of woodland vegetation at the land-sea interface is essential for the persistence of the reef since it filters out heavy silt inputs which would choke the reef.

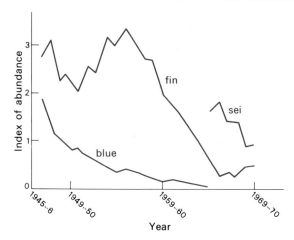

Figure 3.18 An index of abundance (derived mostly from catches per unit of fishing effort) for some of the major species of whales. The curves tell their own story, which reflects little credit on those responsible.

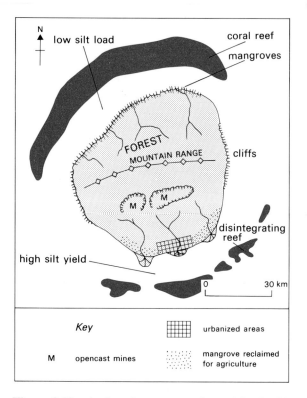

Figure 3.19 An imaginary mountainous island with different land uses on each of the two sides. On the north side the pristine forest remains and the shores are vegetated with mangrove swamps. These hold any run-off silt and so coral–algal reefs flourish offshore. On the south side large quarries contribute a lot of the silt to the runoff and the mangroves have been mostly reclaimed for agriculture or urban growth. The ensuing silt levels have choked much of the coral–algal reef.

growth of phytoplankton and increases the already heavy input of organic matter to the deeper waters in a kind of vicious spiral. Equally important, though not immediately quantifiable, are the effects of reclaiming estuaries, salt marshes and other intertidal communities which house the larval stages of many marine organisms. Best known of all is the ability of modern man to reduce populations of whales and fish to very low levels. Nobody now thinks that the oceans are so vast and so teeming with life that human effects are confined to a narrow zone round the shore, and that they are an inexhaustible supply of food or a bottomless sink for wastes; but putting such realisations into action is proving a slow process.

Perhaps the most fascinating of the high-productivity systems are the coral reefs, found where the water temperature is not less than 21°C. Their physical structure is dominated by the skeletons of corals, which are carnivorous animals living off zooplankton. But their biomass is exceeded three times by the weight of algae living in and around the corals and providing the basic energy fixation of the system, so the proper name of the system is a coral–algal reef. The reef is the habitat for many other animals.

The high productivity ($2500 \ \mathrm{g \ m^{-2} \ yr^{-1}}$ of NPP) is made possible firstly by the flowing of water round the reef which removes wastes and brings in plankton, leading to the second outstanding feature of its metabolism. Thus is its success in holding nutrients (especially phosphorus)

tightly within the ecosystem and recycling them rapidly between algae and corals so that their growth is unimpaired by the general scarcity of phosphorus in the surroundings. The immediate source of phosphorus appears to be the zooplankton, so that one role of the corals is that of a nutrient trap for the algae. No organic matter is permanently locked up in the structure of the reef and so nearly all the phosphorus (and other mineral nutrients) is kept rapidly cycling within the reef ecosystem with the exception of skeleton builders such as calcium. If there are losses to mobile animals, then they are presumably balanced by the gains from zooplankton brought in by water currents.

In spite of its rocky appearance, the coral reef is a relatively fragile ecosystem. Apart from direct

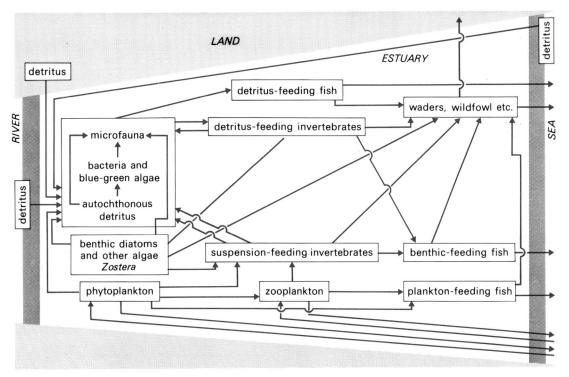

Figure 3.20   A simplified food web for a tidal estuary. Note the inputs of energy and matter in the form of detritus from upstream. Some of the primary productivity (top box of the three representing this trophic level) derives from organisms on the exposed mud flats of the intertidal zone. *Zostera* is a grass of this intertidal zone. In general, phytoplankton production will be low because the water is turbid. The output arrows show several species of fish, emphasising the role of estuaries as nurseries for commercially important species.

damage caused by blasting, construction works or the removal of coral and shells to sell to tourists, certain land-use practices on nearby land masses will affect the reef. Engineering of the river channels which then flush a reef with large quantities of fresh or brackish water is one example, but more common is the deposition of large quantities of silt which choke the reef organisms. Deforestation of a mountainous island will produce such an effect, as will dredging of a lagoonal harbour, and the 'reclamation' of coastal ecosystems which trap silt (of which the mangrove swamp is the most efficient) is usually sufficient to kill the reef (Fig. 3.19). In recent years many coral reefs have been devastated by a population explosion of a predatory starfish, the crown of thorns (*Acanthaster planci*); the reasons for the outbreak are not known, but human activity in some form has been suspected.

A very productive intertidal environment is the estuary with its set of mud flats and salt marshes (Fig. 3.20). With tidal effects, strong currents, high turbidity and variable salinity, the estuary imposes a high degree of stress upon its organisms: a wide tolerance of variability in salinity is necessary, for instance. Plants are salt tolerant (halophytic) and include grass- or rush-like genera such as *Spartina, Salicornia* and *Scirpus,* as well as green algae such as the sea-lettuce *Enteromorpha,* while the mud is covered at low tide with diatoms and blue-green algae. Few phytoplankton are found because of the high turbidity.

There are compensating features: dissolved-oxygen levels are high because of the turbulence; the intermixture of salt and freshwater acts as a nutrient trap, keeping river-borne nutrients in the estuary for a long time in spite of the river current. In the Tropics at least, nutrients come from the sea also and especially from deep waters below the euphotic zone where they have not been depleted by the phytoplankton. The tide does a lot of work in removing wastes and transporting nutrients and organic matter so that the permanent biota can be sessile and do not expend energy on excretion and

Table 3.2 Summary of biome characteristics (for distribution of biomes see Fig. 3.1).

Biome	Typical climate	Structure and flora	Characteristic fauna
tropical evergreen forests	constant high temperature (35°C) > 2000 mm yr^{-1} ppn., at least 120 mm in driest months	evergreen forest with 3 TL; little shrub or ground flora; areas of very high diversity of species, with others dominated by fewer species; epiphytes, climbers and lianes common	stratified like forest, e.g. birds, monkeys; perhaps 60 per cent of mammals arboreal
tropical seasonal forests	2000–2400 mm yr^{-1} ppn., 3–4 months < 25 mm; no cold period below 20°C	1 or 2 TL; shrub layer may be present; deciduous trees may be dominant or mixture of evergreen and deciduous; single-species dominance more common than evergreen forest	stratified forest but fewer arboreal species; habitat of tiger
temperate deciduous forest	600–2250 mm yr^{-1} ppn. evenly distributed; temperature range – 29° to 38°C	1 TL; shrub, herb and moss/lichen layers frequently present; low species diversity may lead to single-species dominance	some stratification of fauna; mammal herbivore density important in regeneration and growth; deer typical large mamma
boreal coniferous forests	350–2500 mm yr^{-1} ppn.; – 56° to 26°C; snow important	1 TL and ground layer; ground layer varies according to moisture, dry lichens to bog; large areas of single-species dominance	definite stratification; many animals migratory; large herbivores may make browse-line in winter; moose and wolf typical large mammals
sclerophyll vegetation	250–850 mm yr^{-1} ppn.; several dry months; 2° to 40°C	low trees and shrubs dominant; sclerophyll adaptation to long dry season	variety of cover and food so species diversity high; deer typica large mammals, bear occasionally
temperate grassland	300–2000 mm yr^{-1} ppn.; – 45° to 45°C	dominants are grasses, but herbs present; tightly knit sod except on dry fringes; roots may penetrate very deeply for water	burrowing animals plus mammal herbivores, e.g. antelope, bison, llama, kangaroo
tundra	250–750 mm yr^{-1} ppn.; snow important in winter	low birches, willows in sheltered places; sedges in wet areas; lichens on very dry ridges	small mammals (e.g. lemming) attract many predators; migratory habit common, e.g. caribou with associated wolf as predator
desert	250 mm yr^{-1} or less ppn.; high diurnal variation in temperature, up to 57°C	drought-resistant shrubs, cacti, succulents and ephemerals in rainy seasons; density of vegetation related to moisture available	burrowing animals, reptiles; some mammals, e.g. antelopes, camels
savanna	250–2000 mm yr^{-1} ppn.; distinct dry season; minimum temperature 18°C; maximum 35°C	trees at variable density; grasses form tight sward	diverse suite of herbivorous mammals and associated predato
islands	very variable; depends on location and altitude	same physiognomy as continents but usually fewer species	poor fauna cf continents; flightless birds common; radial adaptation of 1 group to all habitats

Average NPP $g\,m^{-2}\,yr^{-1}$	World NPP $\times 10^9\,t\,yr^{-1}$	Nutrient cycling	Role of fire	Index of human impact (0–10)
2200	37·4	rapid; very conservative; most of capital in vegetation; no litter accumulation	± absent except at fringes	6
1600	12·0	as above, but slower so litter accumulates	may be natural in dry season if litter accumulated	7
1200	8·4	active soil fauna and mixing with mineral soil; litter layer always present but not thick	not usually significant	9
800	9·6	slow decay of litter, especially where wet; but 1 m of litter possible	highly significant; may mineralise litter layer	5
700	6·0	litter may accumulate to some depth; decay is seasonal	highly significant; helps release minerals to vegetation	6
600	5·4	litter turnover quite fast, but at end of dry season may be deep	often present; consumes plant litter	8
140	1·1	dry areas produce little litter but wet litter accumulates to form peat; nutrient release to vegetation is slow	important on dry areas; vegetation very slow to recover	4
90	1·6	slow decay of organic matter on soil surface	usually insignificant	3
900	13·5	slow decay of grasses, especially in dry season, so considerable litter	highly significant ecological influence in this biome	6
not relevant	not relevant	not relevant	very variable but often found, especially on volcanic islands	8

Photograph 21   The estuary of the River Mawddach in Wales. Such places are highly productive biologically, and also act as nurseries for commercial fisheries. They are also easy to transform into other land uses since the intertidal zone is flat and relatively easily 'reclaimed'.

food gathering. The outcome is an average NPP of 1500 g m^{-2} yr^{-1} (compared with continental shelf figures of 360 g m^{-2} yr^{-1} and open ocean of 126 g m^{-2} yr^{-1}), a quantity similar to tropical seasonal forests. The productivity of the algae and higher plants goes mostly into a decomposer chain in which bacterial breakdown is an important stage, as is the activity of molluscs, worms and other detritus feeders. Bivalve molluscs (e.g. clams, cockles, mussels) in particular exhibit a high productivity: in northern Europe figures of 200 g m^{-2} of dry meat biomass of mussels have been reported. A properly managed mussel bed should provide 2000 kg ha^{-1} yr^{-1} live weight, about 50–100 times the yield from beef cattle on grassland. The visible carnivore fauna, including

the birds which find this ecosystem such an important feeding ground, live off these sessile animals. Importantly for resource use, two-thirds of the commercial fish species of continental shelves spend their larval years in estuaries, and others must pass through them on their way from ocean to upriver spawning ground. The estuaries are therefore important nurseries of commercial fish.

The level terrain of these ecosystems makes them easy to reclaim for many kinds of industrial purpose and estuaries in general are very prone to environmental contamination. Unspectacular though they are, these places deserve a high degree of environmental protection, not merely because of their bird populations but because of their contribution to the whole marine environment short of the open oceans.

3.11   General

It remains to be said that the map of biomes (see Fig. 3.1) is partly conjectural. As was said at the beginning of this chapter, it is largely a map of what the world would be like if man's activities were suddenly removed and all the ensuing successions were telescoped in time; alternatively it might be a map of some time in the past before human manipulation had become significant but after most of the major post-Pleistocene climatic changes and subsequent vegetational adjustments had taken place – perhaps about 1000 BC. But even then, large areas of southwest Asia must have been altered by agriculture and pastoralism and large tracts of southeast Asia converted to cereal growing. It is most certainly not a map of contemporary reality. Most of the tundra, for example, is still there and little altered but most of the deciduous forests of Eurasia have gone and the lowland tropical forests of the Zaire and Amazon basins are shrinking fast. This leads to one more reservation about the reality of the biomes; in the case of those which have been altered by agriculture and pastoralism, research has been done on relict areas either accidentally or deliberately preserved. We do not know exactly to what extent their ecology differs from that of the biome in its pristine state.

Chapter 4

Man and Biogeographical Processes

'Beasts, Birds, Fish, Flowers
Do what the season insists they must,
But Man schedules the Days on
Which he may do what he should.'

(W. H. Auden)

4.1 The different animal

There is no denying that man is an animal. He is subject to the same laws of thermodynamics and obeys the same law of gravity as other animals. He displays the same basic biological characteristics as other animals, and is classified by taxonomists as one of the Primates – the large apes. From the morphological point of view, *Homo sapiens* is one of the largest apes, notable for its upright stance and bipedal locomotion and for being very much less hairy and larger brained than its nearest living relatives. But that is clearly not the whole story, for our species goes beyond these diagnostic characters in its behaviour. We share with some apes a talent for tool making but went far beyond them in our skills at it, even at an early stage of human evolution, when we made tools to make tools. We have developed the arts of communication by the development of language (both verbal and written) in a way only hinted at by ape behaviour and so can pass on ideas and facts in great quantities and with great efficiency. Out of these characteristics has come the development of technology and, later, science. The development of technology started with the evolution of specialised tools and the mastery of fire – ('Man is an animal who cooks his food') – and proceeded through a series of inventions to the use of stored photosynthesis in the form of fossil fuels and to the secrets of the atomic nucleus.

One upshot of this history is intellectual: that is, man is a self-conscious animal in the way others are not. Only we have the characteristics of introspection and of projecting our concerns into the future, and only we have developed a conscious awareness of a moral sense in which some actions are thought to be right and some to be wrong, independently of whether they affect our immediate survival.

Another consequence is more practical – that is, we have learned to manipulate the rest of the world. Some parts of it are too large, like a tropical hurricane for example, to be affected by us, but most living things are not so. They are within our reach and we can affect them in various ways. In the course of human history we have done this a great deal and there are two major categories of effect as far as living organisms are concerned. Firstly, we have changed the **genetics** of plants and animals so that their offspring conform more nearly to our ideas of their utility rather than those of nature's. This process is essentially what we call domestication, in which we have changed the genetic structure of wild plants and animals to the point where they can no longer survive without human intervention. Secondly, we have changed the ecosystems in which the organisms live; in some cases we have altered the physical conditions (as with irrigation) and in others we have changed the species composition or the

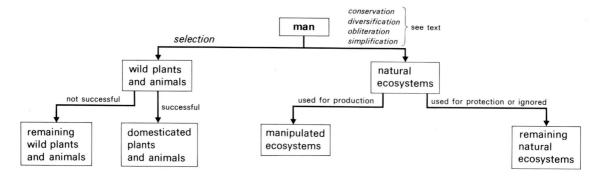

Figure 4.1   A scheme of man's effects on the genetics of wild plants and animals (left-hand side) and on ecosystems (right-hand side). In this context, 'production' means use for agriculture, forestry, etc.; 'protection' means use as nature reserves or extensive recreation areas.

abundance of a species in an ecosystem and thus caused repercussions throughout its component parts. Some of the ways in which we have done this are by simplifying the system by removing some of its species, sometimes to the point of virtually obliterating it altogether by building over it or by poisoning all the life. By contrast sometimes we have created greater biological diversity, not only through domestication but by introducing species from their native habitats to other places. Finally, we have attempted to protect some species and some ecosystems by creating reserves for them, where the processes we have just mentioned are held at bay – the idea of conservation. All these are expanded upon in this section of the book (Fig. 4.1).

The extent to which we have been able, as a species, to effect these changes has been largely dependent upon the development of our technology. This, in turn, can be seen largely as the development of our access to a series of energy sources (Table 4.1).

As we have seen, without access to the energy of the Sun life would not exist – for either plant or animal. But man has learnt not only to tap the energy of the Sun through domesticated plants and animals but to harness energy in other ways as well and to direct the flows of that energy back on to nature and change it in such ways as he wishes. The first of these sources is fire, the control of which our species learned during the Palaeolithic era. Fire confers a great deal of power to alter ecosystems, in particular to clear certain types of forest, to alter the species composition of a grassland or savanna and to affect, for example, the flowering and leafing time of plants; and in some habitats it is invaluable for hunting since wild beasts flee from it.

The next stage was the use of the Sun's energy via that of wind and falling water to aid various agricultural and industrial processes: Medieval wind and water mills are examples of this stage of development and there was even one tide-mill recorded for England in the Domesday Book. None of these, perhaps, had very direct effects upon ecosystems except on-site, although the use of water power to crush ores, for example, may have spread metal smelting in some areas and

Table 4.1   **A table of the energy sources available to man at various cultural levels.** There are still a few hunter–gatherers about, but in general that stage has almost entirely disappeared from the Earth. The 'West' in this context means the whole of Eurasia.

	Fire from wood, other plants, dung	Wind	Water	Domestic animals	Fossil fuels	Nuclear energy	Dates in West
hunter-gatherer	×						Pleistocene
primitive agriculturalist	×			×			9000 BC
advanced agriculturalist	×	×	×	×			1000 BC
industrial era	×	×	×	×	×		1800 AD
nuclear era	×	×	×	×	×	×	1945 AD

thus led to the creation of holes and heaps and to the management of woodland to provide the charcoal needed for smelting ores.

An enormous change came with the discovery of the use of coal, oil and natural gas to power machines, including those which were mobile and self-contained such as the steam locomotive and the steamship. Not only was the power to change ecosystems locally enhanced by thousands of times, but men were able to penetrate further into previously remote places and garner materials cheaply. Eventually they were able to fly and even to land other men on the Moon. An intensification of this process is currently under way with the use of the energy produced by atomic fission to add to our electricity supplies. This Industrial Revolution is the biggest change in the nature of the relationship between man and other living creatures that has ever taken place and we still do not comprehend its full effects, partly because we are creating new ones all the time at an ever faster pace. In ecological terms, this revolution enabled man to add stored energy to that of the Sun, so

mechanising agriculture, for example, and increasing its output considerably. Similarly, the application of industrial fuels has speeded up many nutrient cycles (as with agricultural chemicals) but at the same time opened them up spatially as organic materials are transported around the world. Such power has also brought about the concentration of materials to levels far beyond those found under natural conditions.

To give some life to these generalisations we shall now consider some of the ways in which human activities and biogeographical processes interact, discovering on the way that in many places now they cannot be separated.

4.2 Domestication

The process of domestication has a very long history and its initial accomplishments so changed the relationships between man and nature that they are often called the Neolithic Revolution, after the cultural stage at which it began. The effect of domestication is to replace the processes of natural selection by those of

Photograph 22   A modern fertilizer plant near Calgary, Canada. The biogeographical significance of such a plant is (a) the input of nutrients it makes possible to agricultural ecosystems, and (b) the way in which it ties agriculture to the availability and price of fossil fuels.

Photograph 23 An engraving of the last specimen of *Bos primigenius.*

human selection, in order to produce a plant or animal which is suited to human use. In performing such selection, the aim is to change the genetics of the organism so that the new characteristics are passed down to succeeding generations. Because of this breeding, the domesticated plant or animal is often reproductively isolated from its wild ancestor and so becomes effectively a new species. It is often one which would not survive in the wild: many cultivated cereals for example probably would not survive competition with wild grasses and weedy species; many varieties of pet dogs seem unlikely to have the ability to survive in the wild away from their tins of Wonderchunks. Another part of the process of domestication is often the manipulation of the ecosystem in which the organisms are embedded: competitors are

Figure 4.2   Map (a) shows the dates and locations of the earliest domestications of plants and animals. A large number of domestications seem to have occurred around 7000–6500 BC, with a later surge at 3000–2000 BC. The clustering of dates makes it look as if agriculture was 'invented' independently at different places, not diffused outwards from a single source.

Map (b) shows the places where the major plant domestications occurred. A number of important areas appear and are usually called 'foci'. The 'Near East' centre is sometimes called 'Southwest Asia'. Some crops appear to have been domesticated independently in more than one major focus and these are given in brackets, e.g. grapes, sweet potato. None of the New World domesticates, however, was also domesticated in the Old World since their wild ancestors were endemic to the Americas. Note that the African rice and Oriental rice are different species.

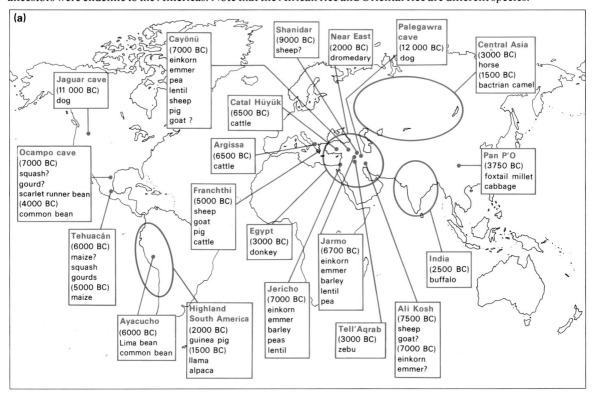

One result is often a distinct change in shape and size of the domestic variety. In plants this is often in the direction of a larger size – if not the whole plant then of the edible parts such as the seed or leaves. Certain other characteristics will also be bred for: in cereals it is vital to breed seed-heads that ripen together and retain the seeds until they are threshed out by human activity rather than spread over the land by natural processes. In animals, one result of domestication often seems to have been a lessening of size; not surprising if we compare the wild cattle (*Bos primigenius*) with today's more tractable-sized beasts. With animals too, certain behavioural modifications can be made in individuals rather than by waiting for the slower processes of breeding. While selection for docility in males may be bred by using the more sedate animals to breed

from, it can also be achieved on a short-term basis by castration. Such animals often put on a lot of fat which was also another desirable characteristic until recent demands for leanness in, for example, bacon. Shorter hair and more matted coats are also side-effects of domestication processes.

The earliest evidence for breeding so far comes from the Nile Valley and is for the cultivation of barley in about 16 000 BC. This is a rather isolated instance (although further research may eventually show it to be otherwise) and the first great surge of domestications are those which we associate with the hill-lands of southwest Asia (modern Iran, Iraq and Palestine) beginning in about 9000 BC and continuing to about 3000 BC. From this 'focus' of domestication we derive animals such as sheep, goats and cattle, and such basic crops as barley, wheat, oats and rye. As Figure 4.2 shows, a number of other plants and animals are associated with this focus as well.

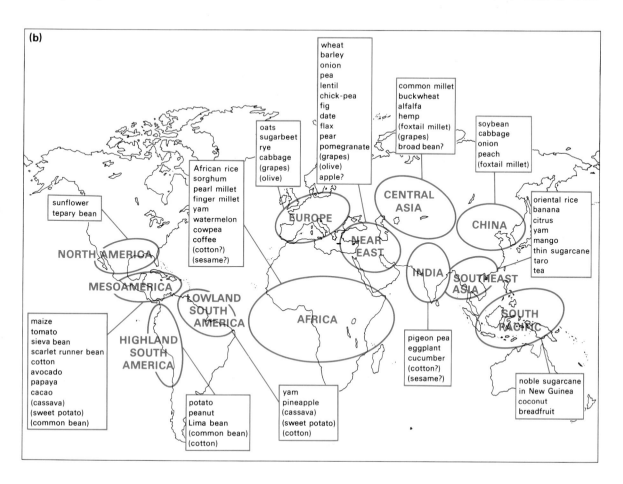

Another ancient focus was southeast Asia which gave us rice and the chicken; and the third major area was meso-America with maize from 5000 BC and other products such as the llama and beans. A few domestications seem to have happened independently in more than one place: the domestication of the pig from its widespread ancestral wild pig (*Sus scrofa*) happened in several parts of the Old World, and the dog (one of the earliest domestications, found from 12 000 BC onwards) is unlikely to have originated from a single centre. Each of these foci seem to have spawned secondary domestications at slightly later dates than the initial surges: the horse came out of Asia in about 3000 BC and the Bactrian camel in 1500 BC: potatoes and tomatoes followed the maize and squashes of the New World.

The process of domestication did not stop with the ancient world and with the establishment of a full post-Neolithic economy. Rather, it is a process which has continued through time – right up to the present day – when scientific knowledge has greatly increased the volume and effectiveness of domestication through the application of plant

and animal breeding programmes. Although the historical process has been uneven (Fig. 4.3), and the last few centuries have seen relatively few totally new domestications from the wild, it seems best to regard domestication as a continuous process, since even the first crops to be domesticated are being improved still. Present-day attempts to domesticate animals include two of the mammal herbivores of North America: the bison (nearly exterminated in the 19th century), which may be able to turn to meat otherwise inedible forage plants of the High Plains of the USA, and the musk-ox, which in its yield of very fine pelt may form the basis for a renewable-resource economy for some of the native peoples of Arctic Canada. We have concentrated, of course, on the major utilitarian organisms, but down the centuries there have been some minor domestications of a rather different kind from the work-a-day cereals and beasts: the Romans, for example, domesticated the dormouse, keeping

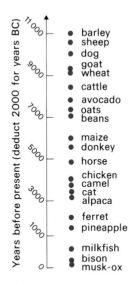

Figure 4.3 Along the axis of years before the present time is plotted the earliest known date of domestication of a selected number of plants and animals. Thus, although the surges noted in Figure 4.2 are of great importance, they are seen as part of a continuing process which is still going on with attempts, for example, to domesticate the musk-ox which are being carried on in Arctic Canada.

Photograph 24 A musk ox *(Ovibos moschatus)* bull. A rare Arctic animal, attempts are now being made to domesticate it, both to ensure its survival and to provide an economy for native peoples of the Canadian North.

century, plant and animal breeders have been able to ensure the outcome of their programmes with increasing success. Much of their effort has been devoted not to taming completely wild species but to improving the yield and efficiency of the basic species already in wide use. In the case of cereals, for instance, breeders have managed to incorporate into varieties with a high yield certain qualities such as resistance to fungal infestation or to lodging (being blown down by the wind). Animal breeders have made specialised versions of species for particular environments, e.g. the hardy hill sheep of the British uplands, or have bred cattle which are especially good at either milk production or meat gain but not both, while a lean variety of pig has superseded the fat-bacon producer of last century (Fig. 4.4). Certain research institutes have specialised in particular crops and it is from these that the well-known high yield varieties (HYVs) of cereals such as rice, maize and wheat were sent out in the 1960s to less developed nations in the hope of curing their nutritional problems – the so-called Green Revolution. In many cases the yields were increased enormously, although there has been a price to pay in terms of the linkage of the Third World economies to industrial world prices and availabilities of fuels, pesticides and fertilizers, as well as some social changes of an unforeseen kind.

The present culmination of plant and animal breeding seems to be in the engineering of the very material of genetic structure itself. It is now possible to 'splice' material from one **gene** on to the genes of another species, and so introduce a new genetic characteristic (Fig. 4.5). It may thus be possible to produce micro-organisms which manufacture substances in short supply – as has

Figure 4.4 Contrasts in pigs. The upper animal is the wild boar which is the ancestral species of all domestic pigs. The lower domesticate is notably larger than its forerunners, having been bred for meat production, and in the course of the years has lost the long hair of the wild animal.

the tame ones in pots of clay to fatten them up for feasting purposes; and the Chinese brought the silk-moth into the human fold during the third millenium BC in order to extend their production of a luxury fabric. In some parts of Japan, tame cormorants are seen fishing in the rivers, where the fact that their catch has been regurgitated into the fisherman's boat does not seem to affect its marketability.

The advent of modern science has put the breeding of plants and animals on a very firm basis of predictability. Ever since the discovery of the basic mechanisms of genetics in the 19th

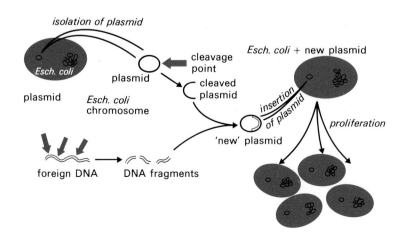

Figure 4.5 A diagram of the basis of genetic manipulation. A plasmid (a circular piece of the genetic material) is extracted from a bacterium, cleaved and then annealed with a piece of 'foreign' genetic material. Inserted into another bacterium of the same species, the new genetic characteristics are exhibited when the organism reproduces. In fact, a new 'species' has been created.

been done with the cancer-suppressing agent interferon, first produced in relatively large quantities by a 'spliced' bacterium in 1980; it may in the future be possible to splice into many species of plant the ability to fix their own nitrogen directly from the atmosphere and thus make them free of expensive and potentially polluting fertilizers. One unusual result of these developments is that it now looks likely that the ownership of a whole **taxon** (i.e. all the members of a particular species created through 'splicing') may be vested in an individual or a firm, rather like any other patent – the first time in history that this has been legally possible although we might recall the activities of orchid-hunters in the 19th century who, having shipped home sufficient specimens for their clients' needs, then destroyed the rest of the natural population to give their employers a monopoly. The application of these techniques of genetic manipulation have so far been benign, but no doubt somewhere the military applications are being given intensive research and development.

If we now turn to man's effect on ecosystems rather than on the genetics of species, we must firstly note a long history of his impact on them. It is not known just when in human prehistory man began to exert an influence upon ecological systems which marked him out from other animals, but we may be certain that it came with the control of fire and perhaps well before that, with the knowledge conferred by language possibly leading to very effective hunting which had more impact than an ordinary predator. As we have seen, the role of fire is very important in the ecology of several biomes and therefore is likely to be of great antiquity. Indeed, it has been suggested that in the savannas of east and central Africa, the vegetation, the suite of savanna animals and early man all evolved together from Tertiary times onwards, with fire as an important element in all their environments. Fire may have had a role in the processes of domestication as well: it has been suggested that the human communities of southwest Asia enlarged the area of grassland at the expense of oak woodland in order to attract firstly wild sheep and goats and then to provide grazing for the ensuing domesticates.

Another long-term process, common in prehistory and still carried out today, is the clearing of forest and grassland for agriculture on a temporary basis – shifting cultivation. It is arguable that this only represents a temporary modification of ecosystems and that there is

Photograph 25   Plant breeding at work. Two varieties of wheat are shown in adjacent plots at the Plant Breeding Institute, Cambridge. The aim is to produce a short-strawed variety and the progress made can be seen by comparing 'Holdfast' with 'Little Joss'.

subsequent recolonisation to the mature stage again, but at any rate in some systems the original forest never returns; it always bears a trace of having been cleared and this difference is emphasised when the cycle of shifting cultivation returns to the same spot once again. In some parts of the world the forest eventually never returns since its clearance produces a soil type which is unable to support forest trees. **Pastoralism** is another very ancient process, although it is likely that it developed after settled agriculture and not as a precursor to it. In this system, domesticated animals are the basis of the nutrition and wealth of the culture. The actual animal varies: cattle, sheep and goats are worldwide and there are some specialities in particular habitats, such as the

Photograph 26   Tropical forest in Peru, cleared for the cultivation of bananas. In this case, the clearing will last a long time but the banana trees will mimic the original forest. If annual crops were being grown, then the clearing would be relatively short-lived and the forest would then recolonise the clearing. In both cases, agriculture represents a considerably more simplified ecosystem than the pre-existing clearing.

llama in highland South America and the yak in the high plateaux of central Asia, and the camel in the semi-arid and desert areas of North Africa and Asia. A herd of animals exerts a strong impact on its forage resources (especially in the marginal lands where pastoralism is most often found) and so the practice of nomadism is often followed. Even so, all domestic animals graze selectively and so the vegetation is gradually altered, as the more palatable plants are eaten by the animals. In areas where the pastoralism is intensive, it is possible for 'over-grazing' to occur and the vegetation becomes more desert-like. Eventually, soil erosion may set in and the area may become reduced to a virtual desert. Thus to be successful and to provide a stable economic system, pastoralism must keep moving – a difficult task in today's conditions of national boundaries and governments with a liking for sedentary and taxable citizens.

All these processes, together with others such as the use of woodlands for fuel and building materials, the changes made in populations of wild beasts which were hunted, the laying waste of

Photograph 27 The pastoralism of domesticated animals (in this case goats in southern Spain) inevitably causes shifts in the vegetation. The dominance of a few species of shrub and the openness of the ground layer may well be due to herding of the goats here over many years.

large areas during times of conflict, were found in pre-industrial times; some of them have become more widespread and intensive since the Industrial Revolution, partly because of the effectiveness of technology and partly because of the large rise in human numbers since the 17th

century. All of them helped to reduce the variety of species found in the original, pre-interference ecosystems.

4.3 Simplification

One of the most widespread methods of simplification of ecosystems in time and space has been agriculture. If we compare the diversity of species in, for example, a woodland with the agriculture that replaced it when the forest was cleared, then the point is made. Again, instead of a complex web of self-maintaining systems, we now have an unstable community which requires much human effort to keep it in the desired condition. To some extent the cultivated field resembles the early successional phases of ecosystems where the food chains are short, where one species replaces another quite quickly and where the nutrient flow is quite open (i.e. there are large losses to the air and to runoff) because the biological mechanisms for a tight cycle have not yet come into being.

The productivity of agriculture is naturally enormously variable since conditions, both natural and social, are different in many places on the globe. The average productivity of agricultural systems for the Earth has been calculated at $650 \text{ g m}^{-2} \text{ yr}^{-1}$, which places it in the same bracket as the temperate grassland biomes and the upwelling zones of the oceans (see Table 2.2). However, it is not clear to what extent this average figure is for agriculture which is powered only by solar energy or whether it represents some fossil fuel input as well. In terms of energy efficiencies, the whole flora of the globe probably uses about 0.2 per cent of the incident radiation of the right wavelengths for NPP, and of that fraction less

Figure 4.6 The 'food system' in a modern economy. The crop captures solar energy (and, not shown, mineral nutrients from the soil). However, in modern agriculture a great deal of fossil fuel-based energy is added to the system. Some of it is 'upstream' from the farm (i.e. to the left in the diagram), some of it is 'downstream' in the form of transport, processing and packing (to the right of the crop box on the diagram).

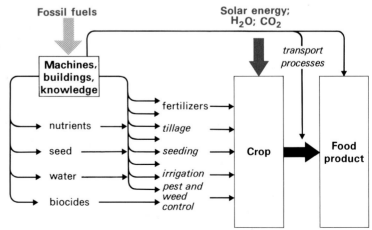

than 1 per cent is used by man as food. So we are barely touching the energy which is available to us if we could but tap it in an acceptable form.

The productivity of agriculture in many parts of the world, and especially in the industrial nations, is boosted by the application of fossil fuels, mostly in the form of petroleum and its derivatives (Fig. 4.6). Machines, fertilizers, biocides and even knowledge are outgrowths of this application. We can get some idea of this if we measure energy density, *viz.* the quantity of energy (solar and fossil) per unit area of an agricultural system. This figure includes both the incoming energy from the Sun and the industrial energy put in by machines and other industrially produced materials. Table 4.2 shows some examples and if we assume for this purpose that the solar input does not vary in space and time (it does, of course, but the differences are small com-

pared with the human-directed inputs), then we can see that there are great differences between agricultural types. Further energy inputs come when food is taken from the farm, transported, processed and perhaps fetched from a shop by car. It is not surprising therefore that some foods when they reach the consumer are energy negative: more energy has gone into their production (in all of its phases) than can be got out of it (Table 4.3). Thus, the simple farms of less developed countries often have a better energy ratio than the complex food systems of the industrial countries, where the food as eaten is dependent on an uninterrupted supply of petroleum – and is said by some people often to taste like it. The fossil fuel input makes possible more storage of food and so buffers us against poor years, and the processing and transport give us much more variety of foods than in a more subsistence-type of

Photograph 28   A mixture of ecosystem types. A semi-natural wood set in a matrix of farmland, most of which is cultivated quite intensively.

Table 4.2   Energy density of food-producing systems (GJ = gigajoules).

Examples	Energy/ density (GJ ha^{-1})	Protein yields (kg ha^{-1})
Andean village (Peru)	0·2	0·5
hill sheep farming (Scotland)	0·6	1–1·5
marginal farming	4	9
open-range beef farm (New Zealand)	5	130
mixed farm in developed country	12–15	300
intensive crop production	15–20	2000
feed-lot animal production	40	300

Table 4.3   Input : output energy ratio for some foods.

Food	$\dfrac{Energy\ out}{Energy\ in} = E_r$
UK cereal farm	1·9
UK dairy farm	0·38
bread (white sliced, wrapped at point of sale)	0·525
broiler hen at farm gate	0·10
winter lettuce in heated glasshouse	0·0023–0·0017
allotment garden produce	1·3
!Kung bushmen (hunter–gatherers)	7·8
shifting cultivation, Congo	65

system. But the latter have survived many stresses over millenia and industrial systems are only about 100 years old – we should not vaunt their advantages too strongly.

The advent of plant breeding together with industrial technology has had one considerable genetic consequence. The requirements for mechanised farms and of industrial food processors have been for a uniform product from the fields. Thus there has been a strong movement towards reducing the number of varieties of agricultural crops in use.

Thus, there are only three major varieties of cotton in use in the USA, constituting about 53 per cent of the crop. Nine varieties of peanuts make up about 95 per cent of their yield. The danger is that one or more of these varieties becomes susceptible to climatic change or to pests and so is lost: there is safety, if not large profits, in diversity. One outcome of this realisation is the building of seed banks and other forms of genetic

conservation. This concern has been evidenced especially in countries to which the Green Revolution has spread, for they have become particularly dependant on large and reliable crops to cope with their fast-growing human populations, and sudden loss to a pest of the whole of a crop which is genetically uniform is more likely than that of a genetically diverse one.

The mention of pests brings us to one of the most common of simplifying processes, that of trying to rid the crop ecosystems of all the plants and animals which are not wanted and which are thus labelled 'weeds' and 'pests'. Pest control in ancient times consisted of simple mechanisms such as children and scarecrows, but has now become a part of the industrial complex, with hundreds of chemical formulations designed to kill other plants, animals, viruses, bacteria and fungi. We ought not to doubt their overall success and the improvement they have made to the steady supply of food in both developed and less developed nations. There have been some side-effects which have created trouble, such as the killing of non-target species (thus reducing diversity still further) and the build-up of resistance in some of the pest species. Also, the killing of some pest species has merely meant that another species moves into its pathways and becomes a pest instead, so that there are a few cases where chemical pest control has really created pests rather than destroyed them. The development of chemicals which kill only the target species, rather than a broad spectrum of organisms may well help with some of these problems. Also becoming more popular slowly is the idea of biological control in which other organisms are brought in to control pests by imitating nature rather than killing them off. Biological control is, however, less predictable and at first more expensive than chemical control and is backed by fewer powerful industrial concerns.

The ultimate loss of diversity through agriculture comes when the agricultural system breaks down completely and soil erosion results. This may happen if there is climatic change perhaps bringing fire conditions, but mostly it is due to the wrong use of the soil – ploughing on too steep a slope, for example, or simply taking too many crops without replacing nutrients or organic matter. Whatever the cause, the incidence of soil erosion of an accelerated kind (not the gradual loss which is inevitable) is widespread in the world and represents the loss of a resource which can only be replaced very slowly – what can be washed away in a single night may take hundreds of years to regenerate (for further discussion see *Soil processes*).

One more simplifier is contamination. Many substances not present in the natural condition which are products of modern industry have been introduced into ecosystems – many synthetic organic chemicals for example. Equally, human activities may be responsible for concentrating natural materials to a degree never found in natural systems: a river's organisms can easily break down the excreta of a group of 25 hunters camped on its banks, but is likely to be overloaded by the waste products of a city of 250 000 with no treatment works. Silt is being carried all the time by rivers and the fauna are adapted to particular levels of it, but a sudden influx from, say, massive clear-cutting in a forest is likely to kill many of the fish who are choked by the extra silt; the same can happen to coral reefs if there is massive deforestation or uncontrolled mining onshore. In any eco-system, a toxic agent normally elicits a differential response, with some species able to tolerate it and others not. Thus there is a biological shift with the result being fewer species than before. Even within a species, the natural variability of a population ensures that there is a difference in response, with some individuals escaping the worst effects to become the parents of the next generation, which are likely to have inherited the resistance; so a resistant population may be built up, as has happened, for example, with certain species of grass on waste tips high in potentially toxic metals. However, the build-up of resistance will vary with the generation time of a species. Thus groups which reproduce rapidly, such as bacteria or certain insects, are more likely to build up resistant populations rapidly than once-a-year breeders.

A well-documented example of simplification by contamination is the effects of air pollution when the atmosphere contains high levels of sulphur dioxide (SO_2) from the combustion of oil or coal. Around big cities and power plants, therefore, there exist rings (Fig. 4.7) of areas where sensitive organisms either cannot grow or else grow only in protected places or hang on rather than flourish. Lichens are very good examples of such plants in cities; several species are intolerant of SO_2 and so they are absent from areas where the fall-out of SO_2 and its solution in water as sulphuric acid is above certain levels. So close is this relationship that some lichens can actually be used as monitors of atmospheric SO_2 levels in the atmosphere. By contrast, the soot component of city air is thought to have been responsible for the spread of a black variety of a moth – *Biston betularia*. This moth showed both pale and black colourations in the 19th century, with the pale predominating, but as time went on the dark form came to be the most often found. It is thought that its dark colour matched the then soot-stained surfaces of trees and fences on which it rested, and thus it more often escaped predation than its lighter siblings. Interestingly, since the Clean Air Act of 1956 in the UK, the paler form is now becoming more frequent again.

Untreated sewage can cause a remarkable loss of diversity near to its outfall into a river, especially at times of low flow and warm water such as summer. The chief agents are bacteria which multiply very rapidly in the conditions of high organic matter, high nitrogen and phosphorus with which they are presented at a sewage outfall. In reproducing so rapidly, they use up nearly all the oxygen in the water, which may be at a low level anyway if the flow is low, causing less mixing, and the temperature is high, decreasing the solubility of oxygen in water. Thus bacteria and a few bottom-dwelling worms such

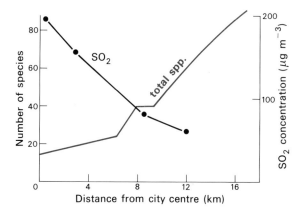

Figure 4.7 The effect of air pollution (defined in terms of sulphur dioxide concentration) on the number of species of lichens and mosses growing on various substrates at different distances west from the centre of Newcastle upon Tyne, UK. As the sulphur dioxide concentration falls, so the number of species rises. Could the number of species be related to other factors, such as the variability and availability of appropriate substrates? We need to know if the sampling has been standardised for such possibilities.

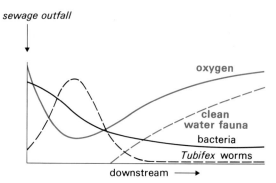

Figure 4.8 The effect of a sewage outfall (untreated material) on various features of the ecology of a river. The effect of the untreated sewage decreases downstream from the outfall. Its most immediate effect is to provide a substrate for bacteria which use up much of the oxygen dissolved in the water. The 'clean water' fauna are replaced by worms of the genus *Tubifex*, which can tolerate the contaminated conditions. A series of similar outfalls would change the ecology of the whole of the river.

Photograph 29 Two specimens of the peppered moth *Biston betularia* **resting on the lichen-covered trunk of a tree.** The upper individual is the melanic (dark) form common in industrial areas, and the lower is the lighter form. In this case, the camouflage of the lighter form is the more effective.

as those of the genus *Tubifex* are likely to be the only organisms in this sort of place (Fig. 4.8).

Overall, it is clear that contamination, as with other influences such as soil erosion, can bring an ecosystem to a highly simplified state, indeed to the point where it has virtually no life at all. This condition, where the system is inert from the point of view of life, we could call obliteration.

4.4 Obliteration

To some extent this may be looked upon as the extension of simplification, but there is the important difference that in the case of the extinction of a species the change is permanent – the genetic unit cannot be recreated. In the case of ecosystems which are obliterated by building, for

instance, or by toxic waste heaps or fallout, the change is in theory reversible, even though it may not occur often with any speed. But abandoned industrial sites, waste tips and even buildings themselves are gradually colonised by some forms of life, and in places such as abandoned tips a succession from lichens and mosses through to scrub can be seen often on tips of different ages. Perhaps we ought not to exaggerate the obliteration of life in such conditions, for nature tenaciously seeks to develop life wherever possible. In the case of cities, there can be found habitat for the roosting of starlings, for example, and food for high populations of pigeons and, increasingly, foxes. Suburban gardens often give cover for numerous bird species and the productivity of town gardens is comparable often with the best agricultural land because of the intensity of labour that goes into it.

The more permanent obliteration is the extinction of a species or other classificatory unit. Extinction, we know from the fossil record, is a normal feature of the evolutionary process: a species becomes unfitted for its environmental conditions and so ceases to grow and reproduce. What has changed is that human activity has the power to accelerate the rate of extinction. We attack species directly and reduce them to such small populations that they cannot survive (as with the dodo and the great auk, for example) or we change their habitat for our own purposes and

Photograph 30   Urbanisation tends to reduce the biomass and productivity of an area, but by variable amounts. In the upper picture, of a typical suburb, there is quite a lot of plant life left in the form of relict and planted trees, and garden plants. In the lower picture, of an industrial area developed in the 19th century, there is very little plant life apparent (except in the churchyard), though there will be lichens mosses, bacteria and algae which are not discernable at this scale.

thus break up the food webs or cover on which they depend. The shrinkage of heathland in southern England has brought a bird like the Dartford Warbler to the verge of elimination from the UK.

There are, of course, spatial scales of extinction. We mostly mean global and total extinction when the term is used, but it can mean local or regional as well. For example, the grazing of sheep in the Scottish Highlands may in the past have rendered extinct several species of arctic-alpine flowers from the UK, but they may well be found still in Scandinavia. At one time, the osprey became extinct in the British Isles due to persecution, but it has now returned from a continental base. Global extinction is more important. In the case of plants, one estimate suggests that some 20 000 plant species are threatened with extinction, which is about 1 in 10 species. (In the USA, this means that somewhere near 750 species out of a flora of 20 000 species are endangered.) Doubtless, among this number there are some which are economically important potentially, as well as those which are scientifically interesting or beautiful, and we need to remember that the interdependence of plants and animals means that a disappearing plant can take with it as many as 10–30 other species of a dependent kind, even including other plants. A group of plants which is especially vulnerable is the endemics of islands; after all, on a continent the chances are that unless a species is very rare a population may survive in some remote or otherwise favourable space. This is not the case with island endemics. We may quote as one example the case of Phillip Island (1000 miles east of Sydney) where introduced grazing animals reduced the forest to scrub and eliminated two species of endemic plants. The destruction of the dodo seems to have stopped the reproduction of the trees whose fruits it ate.

In the case of animals it is similarly possible to suggest the rate at which extinction is occurring. One authority places the rate at about one species of subspecies per year, which compares with about one every ten years in the period 1600–1950 AD (another effect of the Industrial Revolution) and one every 1000 years during the great dying of the dinosaurs. Groups of animals at particular risk include island endemics and colonial animals which are especially vulnerable not only to indiscriminate killing but to introduced predators and mammals with slow reproductive rates at the

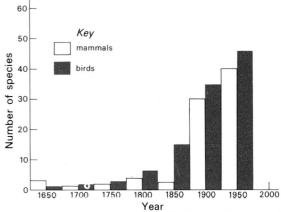

Figure 4.9 An histogram of the extinction of mammals and birds since 1600 AD. The effect of modern economies and population levels scarcely needs emphasis.

top trophic levels. Predators such as the tiger, cheetah and leopard, all of which are valued for their skins as well as feared as the predators of domestic beasts are especially vulnerable (Fig. 4.9).

Efforts to prevent further extinction are dealt with in the section on conservation. Here we will only note that there exists a world list of endangered species, published by the International Union for the Conservation of Nature and Natural Resources (IUCN) and called the *Red data book.*

Naturally enough, there exists as well a list of organisms which man would like to exterminate and we all have our favourite candidates. However, planned extinction has not got very far: the smallpox virus is really the only candidate for this list; even strenuous efforts have not brought about the total elimination of a single species of insect, for example. Heading the list of endangered species through deliberate extinction must, presumably, be ourselves.

4.5 Diversification

Although it is easy to indentify processes such as plant and animal breeding and the introduction of foreign species as adding diversity to the fauna and flora of a place, it must be remembered that they may not add to the diversity of the ecosystems. As we have seen before in this section, they may indeed lead to a diminution in that diversity since they replace all or some of the native species. So diversity must have here a limited meaning.

Breeding of plants and animals for specific purposes has been treated in the section of domestication and will not be repeated here. The bulk of this account deals with introductions. This set of processes has a long history, deriving at first, no doubt, from the simple desire of explorers and other travellers to take reliable sources of nutrition with them as well as perhaps reminders of home; hence, far back into prehistory we have examples of plants and animals being introduced. In the case of the British Isles, one outstanding group of examples is the cereals which were brought to the country by Neolithic immigrants in the years after 3500 BC: though they grew here, the wild ancestors of the plants came from southwest Asia, as did the sheep and goats that were introduced also in prehistoric times. But beyond nutrition, the reasons for which plants and animals were transported, often long distances, are many; economic gain is primary but ornament and pleasure are not to be overlooked, and some have been brought in to act as predators in various forms of biological control. Then, of course, we have the introductions which are accidental. Some of these are purely inadvertent – as when the seeds of a plant are transported in a ship's ballast and others are escapes – a population of wild porcupines is now established in south Devon from a pair which got out from a wildlife park.

There are always risks with introductions. At one end of the scale there is the possibility that the new species will not thrive at all in its new home and will require enormous effort to give it special conditions which will keep it alive; at the other extreme is the species which finds life entirely congenial and becomes so common as to be a pest in its own right. In the first case we might quote the case of aristocratic orchid-collectors from the 19th century in Europe who spared no expense to pamper their hothouse blooms once they had been brought back from the Tropics; in the second the European Sparrow in North America which, introduced into New York to control aphids, soon became a nuisance in its own right – a story not very different from that of the starling on the same continent (Fig. 4.10).

In a few cases, the new species puts a distinct stamp on its new home, to the point where it is naturalised and is often thought of as an indigenous species. The eucalyptus tree, for

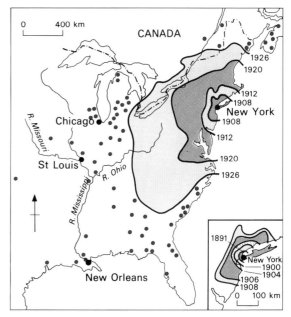

Figure 4.10 The early spread of the European starling *(Sternus vulgaris)* **in North America following its release in 1891 in Central Park, New York City.** Outside the 1926 isochrone, the dots are instances of early phases of colonisation. The starling is now well established in most states and in Canada.

example, is such a familiar sight (and smell) in California that it is difficult to imagine that it has been there only since the 19th century. The case of the rabbit in Australia is well known as a disastrous invader of grazing lands, as is the introduced Prickly Pear cactus, and in New Zealand (a pair of islands with a rather low number of species) there has been a series of instabilities resulting from introduced species. The first came after Captain Cook, who introduced pigs and goats which became wild and began to affect the regeneration of the forests which then covered about 68 per cent of the islands. Deer were also introduced with the European colonists and they exacerbated this process. In 1858 the possum was brought in from Australia to be a fur-producing species; escapees not only defoliated the remaining forests of native trees but have moved on to plantations of introduced conifers as well. Wholesale changes in landscape and ecology have thus resulted from introduced plants and animals.

If Britain is a good sample, then the rate of introductions, like many other processes, accelerated in scale in the 19th century. It was then, for instance, that many of the explorers of North America started to bring back viable seed of the west coast conifers which they thought might well suit British conditions. Many such species have remained arboretal curiosities but the Sitka spruce (*Picea sitchensis*) has become a mainstay of the Forestry Commission's planting programmes since the 1920s, and another, the lodgepole pine (*Pinus contorta*), is also important commercially. The introduction of animals seems to have lagged a little behind that of the trees, for here the years 1850–1950 seem to have been the peak. Few, however, of our recently introduced animals have achieved such an economic status as the Sitka spruce, although the coypu may have done so in a negative kind of way, and the grey squirrel has few friends. But the more significant rabbit, pheasant and black rat are all 12th century immigrants to the British Isles (Fig. 4.11).

A particular feature of the present situation is the pet trade. Increased affluence, mobility and urban life-styles have all combined to produce a flourishing trade in exotic animals. Escapees are common, and some species are got rid of down the w.c. Although most of them doubtless die, there have come to be flourishing colonies of some ex-pet species in foreign places (not to mention the giant alligators said to inhabit the sewers of New York). In the tropical waters of Florida, about 42 non-native species of fish have been found of which 24 have become well established and five interbreed with native species. African tropical fish have become established sometimes in the hot water plumes from power stations in Europe. In Florida again, the hamster and the gerbil are found wild in such numbers as to be thought of as agricultural pests.

Such accounts would be probably no more than amusing were it not for the fact that introduced species tend to bring their pests and parasites with them and they may also spread into native species, creating problems where none existed before. In the case of Britain, the current fear is

Photograph 31 An example of an introduced animal species: the fallow deer (*Dama dama*), probably brought to Britain in Roman times and now established in the wild in several parts of the country.

Photograph 32 An example of an introduced plant species: the giant redwood (*Sequoiadendron giganteum*), endemic to California, growing well in Great Britain.

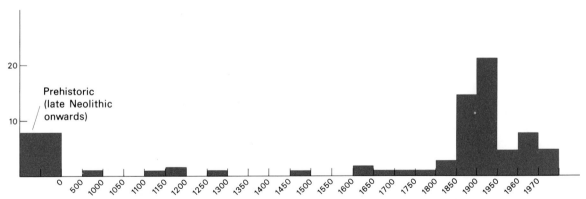

Figure 4.11 Dates and numbers of animal species introduced into Britain and now naturalised. Up to 1600, the naturalisations (i.e. those imported animals which become able to survive in the wild) are rather sporadic, but thereafter rise through the 20th century, apparently falling off in more recent years. The three species established after 1970 are the crested porcupine, Mongolian gerbil and tesselated snake.

that the introduction of mammal pets might be one immediate source of rabies which is present in continental Europe but not in the British Isles.

In general, the movement of people and materials is paralleled by that of plants and animals and so at present something of an homogenisation is taking place, particularly in the spread of uncontrollable organisms such as bacteria and viruses carried by people. It may well be that in a few decades' time we shall have to redraw the maps of floral and faunal regions to reflect a new and man-made reality.

4.6 Conservation

The theme of much of the previous material in this chapter has been man-made change. In the course of these processes there has arisen the human desire to protect some species and some ecosystems from such alteration and to preserve them, if possible in perpetuity. The attempt to preserve species of animals against certain kinds of ecological and economic change is not new, although like so many other things it has gained great impetus since the 19th century. But we can find examples of nature reserves in Asia in Medieval times, for instance, where certain animals were protected for the chase, and the Ancient Greeks preserved sacred groves where the spirit of the place was said to live. So nature and landscape protection have a long history.

Today's motivations for the protection of species and for the protection of landscapes against certain kinds of man-made change (usually those connected with economic activity)

fall into a number of groups. At any one place these may overlap and indeed be implicit rather than explicit, but we can usually find one or more of these reasons in the efforts by both private and government bodies to bring about conservation. The first set of reasons stems from an ethical base and asserts that wild species have a right to co-exist with us on the planet – that we have no right to get rid of them all in the course of our activities and our lives. As might be expected, such an abstract argument often fails to carry the day in the face of immediately practical concerns or in the face of the money waved about by a transnational mineral company, but these are not reasons for discarding the arguments, which have a very serious dimension. A next set of purposes is based upon scientific research. This is partly for its own sake, like all pure science, but is also based on the idea that the study of natural ecosystems may yield very useful information. It is argued, for example, that natural ecosystems are much more efficient at using the incident energy than are man-made ones and so we would do well to study natural processes in order to imitate them when organising our own productive systems. Bacon's famous phrase 'we must obey nature in order to command her' rather sums up this attitude. The science-based view also extends to a rather less well-defined concept in the search for 'eco-system fitness', i.e. what sort of ecosystem is best for human life? Is it one for example in which we mostly live in cities? Is some contact with the natural environment necessary for healthy lives? Do natural environments play an important role in our lives even though we may not go into them very much, just by being there? If we do away with all wild places and wild creatures then we shall never find the answers to these questions.

Scientific arguments are also at the core of the realisation that our plant and animal breeding processes are reducing the amount of genetic diversity, as has been discussed before. Part of the answer is to set up seed and sperm banks and other special places for the preservation of numerous varieties of animals and plants: research stations, botanic gardens, zoos, all have their role to play. But it is realised that firstly, nature will do much of this at low cost if just left alone, as far as wild organisms are concerned, and secondly that in nature there still exists a pool of potentially useful plants and animals whose value to us is not yet realised. For instance, the full range and diversity of the flora and fauna of tropical forests are not yet known to science and there may be many useful plants we do not yet know about. So on the grounds of future usefulness in terms of new organisms for domestication and in terms of new genetic material for old domesticates, there are good reasons for protection of wild organisms and places.

Scientists have also put forward the hypothesis that in some ways ecological stability is related to diversity of species. Initially this argument derived from ecosystem studies where complex food webs ensured that if one species disappeared, there were numerous alternative food supplies for its predators and so they did not become extinct as well. Further, a wide range of predators ensured that one species did not become over-abundant. At one time, the diversity-stability view took on the role of a dogma, but it is now realised that it must be too simplistic a view, for there are many ecosystems which are quite simple but which seem to be stable, just as there are complex ones which can be destabilised quite quickly by the irruption of one species in the way the Crown of Thorns starfish decimated many coral reefs. But the examples of agriculture which lacks genetic variety and the erodability of simple ecosystems such as grouse moors may leave us with the suspicion that although the argument may not be universal, there is a good case for retaining diversity whenever it is currently found, and of increasing it wherever the lack of it is man-induced.

A last set of motivations is subjective in nature, though we may not reject them simply for that reason. There is no doubt that people in many cultures gain pleasure, enjoyment and refreshment from the presence of wild creatures and wild places. This applies to those who watch birds or feed them in their gardens just as much as to those who drive into the countryside at weekends or who take wilderness trips into wild places. Some writers have gone so far as to assert that these activities prove the need for nature as part of the ecosystem fitness argument (see above) but even without going that far, we can assent to the idea that our lives are often enriched by having access to wild plants and animals and to natural environments. The uses of the wild for recreation may, of course, bring conflicts (see below) but may also link the preservation of wild places to the economically powerful demand for recreation and thus ensure a measure of protection which they might not otherwise get.

If we put all these reasons together, we can perhaps see a linking thread to them: it is that evolution by natural selection produces a great variety of living forms and systems. Change by human selection in most respects tends to diminish that diversity. Conservation is at present a swimming against that tide in an effort to

Photograph 33   The rare osprey in Scotland. Common before the 19th century, it was persecuted because of its alleged appetite for the salmon which were one of the sporting species of the lairds' estates. In the more pro-preservation atmosphere of post-1945 Britain, it has again become a breeding species in Scotland and may even be spreading to northern England.

Photograph 34   A protected environment. Part of the Serengeti National Park in Tanzania, with a herd of giraffes among the *Acacia* trees, the tops of which are their chief food.

diminish the rate of attrition of the diversity: its aspiration is to point out a relationship of man and nature which does not even try to take away any of the diversity which nature has brought into being.

A discussion of the ways in which nature is protected can be divided into two parts for convenience, a division which does not always, of course, reflect the more complex state of reality. Firstly, we have the protection of individual species, which are rare perhaps or typical or even sacred. Secondly, we have the protection of ecosystems themselves; some of these may cover large areas and may need to be manipulated in some way in order to maintain their characteristics. This process is called management.

The protection of species may begin and end with legislation which forbids, for example, the picking of a plant or the killing of an animal; this law may extend to the whole of the legislative area

covered. So if a protected hawk starts to take the fish from my garden pond, I may not kill the hawk: if by chance a rare plant takes root in the same place, I may not pluck it. This type of legislation is often used to protect organisms which range over a wide set of habitats, including those which bring them close to human habitation – migratory birds are a good example. However, such attempts at protection may be insufficient and it becomes necessary to establish a refuge for the plant or animal where, it is hoped, it can live unmolested and where its food supplies and other environmental conditions are assured. Such areas are called nature reserves, wildlife refuges or some variant on such terms. They may be set up by national governments, local governments, voluntary bodies or even by individuals.

The protection of an ecosystem, on the other hand, may come about simply because it is remote and far from human influence. Such places are becoming increasingly rare but there are parts of the boreal forests, of the tundra, the deserts and the oceans, for instance, where this is the case and the whole system is in its natural condition. In general, those systems which have remained pristine have done so because they are marginal to human material needs or are unattractive for recreation.

Where there is a need to protect ecosystems within the ambit of human activity then, as with the individual species, some form of legislation is needed which sets the terrain aside from other human activities. Such areas are often larger than the nature reserves mentioned above and very often carry designations such as Game Reserve (especially if the main element is an assemblage of mammals and their predators) or National Park. The term National Park may also be used to denote areas of outstanding scenery which it is desired to protect. Examples are to be found in the famous game reserves of Africa such as the Serengeti, the National Parks of North America (including the US National Wilderness Preservation System) and in the largest protected area of all, Antarctica, which is preserved under the terms of an international treaty.

The motivation for both large and small reserves, dedicated both to individual species and to regional ecosystems, is scarcely different, it is the scale of the protected areas which may vary. This is not always the case of course, for sometimes one species may require a great deal of space. The California condor, of which only a small number remain, seems to be unable to survive at all in the presence of people, so that a large area of the Coast Range is set aside as a refuge for these shy birds and all other activities are subordinated to that purpose. But whatever the scale, both types of area may well have in common the need for active human intervention to carry out the protective purposes of the area: only a few reserves are large enough and remote enough to be left alone with a complete 'hands off' policy. This intervention is called management (Fig. 4.12). At its simplest level, this may mean a fence which deters intruders or notices which draw attention to the status of the area and the need not to disturb it. Naturally, these may be insufficient (and in some places may actually ensure damage) and so guardians, usually called wardens or rangers, are appointed to run the reserve.

Such simple measures may, again, be in-

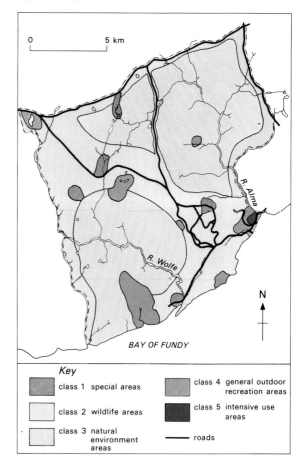

Key

class 1 special areas

class 2 wildlife areas

class 3 natural environment areas

class 4 general outdoor recreation areas

class 5 intensive use areas

—— roads

Figure 4.12 A zoning plan for a Canadian National Park in New Brunswick, showing the areas primarily devoted to various functions. Class 1 special areas, for example, contain rare species or exceptional scenery and so merit especially stringent protection. Class 5 areas are intended to attract most of the casual recreationists and also to contain the facilities needed for visitors, such as campgrounds, toilets and car parks. The road layout is critical in the location of Class 4 and Class 5 areas.

sufficient for if the reserve is surrounded by more intensively used land, then it is likely that the influences of those more manipulated ecosystems will penetrate the reserve. For example, if a reserve is surrounded by farmland then it is quite probable that any runoff which passes through the reserve's waterflow system will carry residual fertilizers (which may enrich its waters) or residual biocides (which may kill or sublethally affect its animals); equally, if the reserve is adjacent to grazing lands, then it may provide a home for a predatory species which is part of

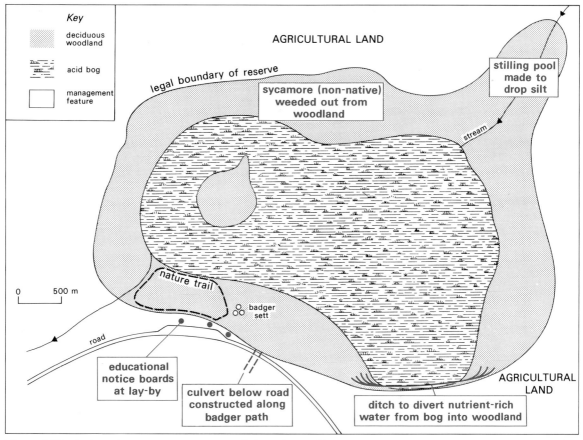

Figure 4.13   An imaginary nature reserve consisting of an acid bog surrounded by deciduous woodland. Around the reserve is agricultural land and it abuts on to a road. Shown in boxes are some of the management actions that might be taken to keep the woodland in its native condition and to prevent runoff of fertilizers and pesticides from the farmland affecting the ecology of the bog. Some limited measure of public access and education is envisaged.

the reserve's natural ecosystem but not welcome on the grazier's manipulated system. A reserve in which the deer were predated upon by wolves would be an instance: from time to time the wolves might fancy a fat sheep rather than a scraggy deer but get shot or poisoned in the attempt. If this considerably reduced the level of wolves, then active intervention by the reserve managers would be needed to prevent the deer population becoming too high for the carrying capacity of the reserve. An analogous example was reported from Africa where elephants started to congregate in certain parks because of agricultural extension elsewhere. Because of the feeding habits of the elephants, they began to uproot so many trees that the feeding patterns of many other animals were disrupted: thus many elephants had to be shot as a management device.

It will become apparent that what is needed for

most reserves is a management plan. The first essential of such a plan is a statement of purpose: what is the reserve to be managed for? The long-term aims of the managers must be clearly stated here. After this, a great deal of scientific knowledge is needed in order to formulate strategies for achieving this aim: knowledge not only of the actual species composition of the reserve, but of the dynamic interactions between them and their non-living environment – in short, a comprehensive knowledge of the ecology of the area in all its aspects. It scarcely needs to be said that this is not always possible. Nevertheless, certain techniques are often open to managers in their attempts to find ways of managing the ecosystems. Where outside influences are causing problems, for instance, it may be feasible to buffer the area: in the case of the agricultural runoff cited above it might be possible to have a ring-ditch which led this water away from the reserve without damaging the water supply of the plants within the area (Fig. 4.13). Plant and animal populations can often be manipulated: animal populations can be culled by shooting or using biocides, for example, (paradoxical though it sounds) and plant populations likewise can be changed. Areas of interesting flora which grew up under conditions of sheep grazing which no longer exist, for example, are prone to invasion by thorn scrub. Volunteer labour is often then used to clear away the bushes which are shading out the herbaceous plants of interest. In some places, apparently extreme measures have been used: ponds which have colonised with floating vegetation to the point of having no more open water and thus losing their duck populations have had an earlier stage of succession restored using dynamite. The giant Sierra redwood tree (*Sequoiadendron giganteum*) of the Sierra Nevada of California does not regenerate in the absence of fire, for rapid ground fires destroy the seedlings of its competitors which, without the fires, rapidly outshade the redwood. However, in a National Park it is scarcely possible to have natural fires, because of the risk to visitors, so instead the management arrange for controlled burning of the area from time to time, organised by their own staff.

Even in a remote place such as Antarctica, many precautionary measures are taken which might be regarded as management. No flora and fauna from the outside world are permitted, for example, in case they explode into the biologically impoverished habitats at the edge of the ice; this applies even to the parasites on the husky dogs. International permission has to be obtained to take specimens of many plants and animals and agreement has had to be reached with cruise operators to avoid the biologically sensitive areas which were beginning to show signs of damage from tourism.

It is interesting, if not indeed ironic, that one of the most potent forces for damage within a designated reserve is the visitor population: above a certain level visitors very often destroy or downgrade the very features that they value and have come to see. Some special reserves have a complete prohibition on visitors but such policies are expensive to organise and enforce, and are to some extent bad publicity for the conservation cause. But as with so many human impacts on nature, what is tolerable with 6 or 10 people twice

a year becomes a different matter when it is 10 000 per day for half the year. At a low level, visitors may do little other than wear a narrow trail but high numbers will turn this trail into a kind of rural freeway, with the wear zone extending outwards rapidly (Fig. 4.14). On soft ground such as mires, much damage may be done quickly and a walkway above the surface of the bog is needed. Low numbers of people may simply frighten an animal temporarily and it returns to its normal behaviour after they have gone. High intensities of people may cause the animal to leave that area altogether, thus diminishing the range over which it is found or, if it can adapt to human food surpluses and cast-offs, then it may become a scavenger off the visitor population. This may

Photograph 35   Environmental alteration for outdoor recreation. This chair-lift not only necessitates cutting down a wide swathe of forest but causes considerable concentrations of people at either end who trample the vegetation and increase the fire risk, to say nothing of the buildings to house the lift-gear. This example is in the Daisetsuzan National Park, Hokkaido, Japan.

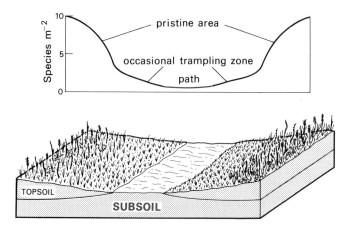

Figure 4.14 Block diagram and graph to show the effect of trampling along a path used intensively for outdoor recreation. The graph shows the dramatic drop in the number of plant species able to tolerate the trampling. Compaction of the soil also means a decrease in the activity of the soil fauna and flora, and less infiltration of precipitation. Thus erosion of the soil and subsoil becomes a possibility.

bring attractive animals such as deer and blue jays to North American campsites, to quote one example, but it may also bring the rather less attractive (when adult) brown bear and even in a few places the grizzly.

Low numbers of visitors may, as the saying has it, 'leave nothing but their footprints and take nothing but photographs' but this is scarcely realistic above. certain levels. A few people urinating behind the bushes is not much of a problem, but 10 000 people require toilets; campers must be supplied with water and waste disposal facilities and soon all the markings of a 'campurbia' are present, especially given current fashions in recreation hardware such as the trail bike and the motor caravan. Soon too, there will be pressure to use a lake or two for water skiing and maybe to install some ski lifts for winter use and then the area will be an outdoor recreation area rather than a nature reserve. Sometimes it is possible to attack this problem by zoning, in which part of a National Park is allowed to have developed facilities for recreation, while the rest is maintained in a pristine condition. In others, changes in management policy have reversed trends of the last 30 years and started to 'de-develop' such places. Public opinion is very often slow to recognise the major purpose of nature reserves and National Parks, thinking that they are primarily for recreation, which is rarely the case.

What is the overall aim, if any, of conservation? It must be to preserve a certain proportion of the Earth's surface in mature ecological systems as distinct from the secondary successions produced by human activity. We do not know if there is a 'right' proportion, although we can be reasonably sure that the present proportion of 1.1 per cent is too low, with many biologically interesting areas not represented at all. Beyond this, and in areas not designated specifically as reserves, the aim of conservation must be to preserve the natural diversity of the living things and their ecosystems on the face of the Earth. To do this properly we must (*pace* W. H. Auden, p. 65) schedule rather more Days than at present.

Chapter 5

Envoi

'I suggest that the Earth's biota is our single
most important resource'

(G. M. Woodwell 1974)

5.1 So what is biogeography?

Even though this book is highly selective in its
material, we have seen that there is quite a
diversity of approaches. The distribution of
species, the nature of ecosystems and the effect of
man on the genetics of organisms, as well as his
role in creating new ecosystems, are all part of the
contents. If we regard the study of natural distributions and that of natural ecosystems as being
part of an attempt to build up a datum line (i.e.
what is nature like without human influence?)
then the rest of the study is concerned with man's
effect on genotypes and man's effect on ecosystems. Here perhaps we have the fundamentals
of a distinctly geographical approach to the study
of plants and animals, one in which we use much
of the information generated by biologists but
look at it in a different way. Our way lies at the
intersection of the natural world and the human
world in this case, and in that way we link biogeography to one of the distinct traditions of study
within geography as a whole – that of man–
environment relations. This strand within our
subject has a long and honourable history and can
scarcely be derided as irrelevant in today's world.

5.2 The real world

Just as important, if not more so, than intellectual
traditions, is the status of plants and animals in
the real world outside our schools and colleges.
First of all, as the quotation by the North
American ecologist G. M. Woodwell at the head
of this chapter suggests, the plants and animals
(collectively called **biota**) are fundamental resources for man. We eat them, raw, cooked and
processed; we make fibres from them; we use
them as flavourings and for perfumes and scents;
we make clothing from animal skins; we use their
fibres for paper and for timber; we get drugs from
them and we use them for experimental purposes
in the development of pharmaceuticals and in
other medical research. This list is not exclusive,
more categories can be added. But all these uses of
biota as resources have one thing in common
which some other kinds of resources do not have
– they are, if managed properly, infinitely renew-
able. The stocks of coal, of oil, of iron ore and of
land surface are all finite, that is, there is a fixed
amount of them and the only way to make them
last is either to use them slowly or to use them
over and over again, i.e. to recycle them. This is
not possible with infnite energy sources of course,
for the stored energy in fossil fuels becomes low-
grade heat which is radiated back into space. But
biological resources are self-renewing because of
their reproductive powers and, if not over-used
and if their environmental conditions are maintained, they will go on yielding useful resources
for as long into the future as we can imagine. This
makes living things a very special category of
resources which we ought perhaps to value far
more highly than we do: they are, as one
geographer has recently put it, 'the real wealth of
nations'.

Biota are, too, what we might call non-
resources, or at any rate resources of a different
kind from those of the last paragraph. Here we
mean the many ways in which plants and animals
contribute to delight and wonder in our lives.
They may do this directly – as in the sight of a herd
of wildebeeste on the plains of Africa or in the
diving of a gannet into the sea, to say nothing of

the purring of a domestic cat – or it may be indirect: in many cases the most valued scenery has a strong biological component in it, in the form of forests, for example, or moorland, or alpine meadows. Also there is the interest in knowing that scientific discoveries about the ecology of the world are still to be made: some of these may turn out to be useful but others no less important may be simply parts of the process of discovery about our planet which makes us yet again a different kind of animal.

It is apparent from what we see about us and from what we read that there are different kinds of relationship between man and the biota of the world. There is firstly a set of relationships which results in the continuation and expansion of the processes of simplification and obliteration. No more examples need be given, for it is a condition with which we are all familiar and which we often may see in action. But there is an alternative viewpoint which stresses the type of attitudes which underlie conservation. If we interpret this word rather widely, it can mean more than just the setting up and management of biotic reserves. It can mean an attitude towards our relationships with nature in which there is, for instance, room for the wild as well as the domesticated, a world with a high diversity of species in it, with none lost if at all possible (and with a high diversity of human cultures as well, we might hope), and in which not everything living is subordinated to the material demands (as distinct from needs) of the human species.

If we wish to shift towards this latter set of attitudes, then certain changes in our behaviour will be necessary. Let us take two fundamental examples which are easy to describe but difficult to put into practice. The first is a revision of our attitudes towards material possessions and use. No conservationist is advocating a return to a lost 'Golden Age' of happy laughing shepherds and their lasses dancing round the village green. Those times never existed: for the lads probably all had rheumatism from an early age and the girls tuberculosis, and neither are conducive to dancing. But conservationists are in favour of reducing material consumption and of using less energy in unnecessary ways, such as planned obsolescence of goods, using large cars for short trips with one person aboard, emitting poisonous wastes into air and water; in all not to deprive people of the things which have made possible in the West an abundance of life never before

achieved, but to make people content with a modest sufficiency rather than an excess. There is evidence too that such a shift might make us happier: less psychic strains from the 'rat-race', less poor health from too much animal fat, too little exercise and too little dietary fibre and fewer workers in environmentally risky industries. This is just the West: the major question to be asked about the less developed countries concerns their pathway towards better standards of living: is it to be a repetition of the West's path, with both its advantages and its problems, or is there another way? There is no simple nor single answer to this but we may note that different approaches to 'development' based on 'soft' or 'alternative' technology (both of which have far less impact on biota than 'heavy' technology) are becoming more popular.

But a biological idea which applies to us all is that of carrying capacity (p. 25). Every ecosystem has its carrying capacity for each species. That is, there is a maximum number of individuals of each species which the energy flows and the nutrient cycles will support. If populations get beyond that level, then feedback mechanisms reduce them again – by starvation, disease, social pathology. The world ecosystem must have a carrying capacity for the human species too. We cannot say what it is. It might be possible to house billions and billions of people in a 2000-storey building covering the whole surface of the Earth and maintained on industrially produced foods. On the other hand, it might be preferable to house rather fewer in a world in which the natural processes still had a large place and in which travel, for example, still had some point. Conservationists see the latter as preferable but point out that the carrying capacity is lower. So the Earth is perhaps like a large nature reserve in one way: it is necessary to have a purpose built into the first paragraph of the management plan. Do we want a world which is largely man-made and in which all nature's mechanisms are replaced by our own? Or do we want one in which nature performs for us all sorts of services free of charge and gives us delight and wonder into the bargain? If we want the latter, then the carrying capacity is lower and so our first task, in industrial and non-industrial nations alike, is to realise that our population dynamics (p. 26), like any other species, must be brought into an equilibrium state as soon as feasible. This dynamic is as fundamental a piece of biogeography as any.

Further Reading

Distributional processes

Cox, C. B., and P. D. Moore, 1980. *Biogeography: an ecological and evolutionary approach,* 3rd edn. Oxford: Blackwell Scientific.

Darlington, P. J. 1957. *Zoogeography: the geographical distribution of animals.* New York: Wiley.

Good, R. 1974. *The geography of the flowering plants,* 4th edn. London: Longman.

MacArthur, R. H. 1972. *Geographical ecology.* New York: Harper and Row.

Ecosystem processes

Lieth, H., and R. H. Whittaker (eds) 1975. *Primary production of the biosphere.* Berlin: Springer.

Odum, E. P. 1971. *Fundamentals of ecology,* 3rd edn. Philadelphia: W. B. Saunders.

Odum, E. P. 1975. *Ecology,* 2nd edn. New York: Holt, Rinehart and Winston.

Scientific American 1970. *The biosphere.* San Francisco: W. H. Freeman.

Biome processes

Collinson, A. S. 1977. *Introduction to world vegetation.* London: George Allen and Unwin.

Eyre, S. R. 1968. *Vegetation and soils: a world picture,* 2nd edn. London: Edward Arnold.

Odum, E. P. 1971. *op. cit.*

Polunin, N. 1960. *Introduction to plant geography.* London: Longman.

Man and biogeographical processes

Bennett, C. F. 1975. *Man and Earth's ecosystems.* New York: Wiley.

Cushing, D. H., and J. J. Walsh (eds) 1976. *The ecology of the seas.* Oxford: Blackwell Scientific.

Davidson, J., and R. Lloyd (eds) 1977. *Conservation and agriculture.* Chichester: Wiley.

Earl, D. E. 1975. *Forest energy and economic development.* Oxford: Clarendon Press.

Eyre, S. R. 1978. *The real wealth of nations.* London: Edward Arnold.

Isaac, E. 1970. *The geography of domestication.* Englewood Cliffs, NJ: Prentice-Hall.

Lenihan, J., and W. W. Fletcher (eds) 1975. *Food, agriculture and the environment.* Glasgow and London: Blackie.

Myers, N. 1972. *The long African day.* New York: Macmillan.

Myers, N. 1979. *The sinking ark.* Oxford: Pergamon Press.

Odum, H. T., and E. P. Odum, 1976. *Energy basis for man and nature.* New York: McGraw-Hill.

Pirie, N. 1969. *Food resources: conventional and novel,* 2nd edn. Harmondsworth: Penguin.

Sauer, C. O. 1962. *Agricultural origins and dispersals,* 2nd edn. Cambridge, Mass: MIT Press.

Schumacher, E. M. 1973. *Small is beautiful.* London: Blond and Briggs.

Sheail, J. 1976. *Nature in trust.* Glasgow and London: Blackie.

Simmons, I. G. 1979. *Biogeography: natural and cultural.* London: Edward Arnold.

Simmons, I. G. 1981. *The ecology of natural resources,* 2nd edn. London: Edward Arnold.

Ziswiler, V. 1967. *Extinct and vanishing animals.* New York: Springer.

Glossary

abiotic Non-living, i.e. a physical or chemical feature of an ecosystem.

aseasonal Any biological process which does not happen at the same time every year, e.g. leaf-fall in lowland equatorial forests.

behaviour Any action taken by an organism to adjust to environmental circumstances.

biogeochemical cycle The movement of chemical elements through living organisms and their non-living environment.

biological productivity The growth of organic material per unit area per unit time. Primary productivity is that of plants, secondary productivity that of animals. It is usually expressed as the dry weight of the organic matter, e.g. g m^{-2} yr^{-1} or t ha^{-1} yr^{-1}.

biomass Total weight of living organisms per unit area.

biomes Major regional communities of plants, animals and soils, covering large areas, e.g. tropical rain forest, tundra.

biosphere The surface of the Earth and the lower part of the atmosphere in which organic life exists.

biota All the living organisms found in a region.

calcicole A plant which grows best on soils rich in calcium salts.

calorific value The energy content of an amount of organic material.

carbohydrates Compounds of carbon, hydrogen and oxygen with the general formula $C_x (H_2O)_y$, e.g. sugars, starch, cellulose.

carnivore An animal or even some plants which eat animal tissues.

carrying capacity The number of individual organisms that the resources of an area can support.

chlorophyll The green pigment of plant cells essential for photosynthesis.

climax The final stable stage in a successional series of plant communities in which the dominant species appear to be completely adapted to the prevailing environmental conditions. If climate is the overriding influence, then the term **climatic climax** may be used.

community A group of different organisms occupying a common environment interacting with each other, while remaining comparatively independent of other groups.

competition Competition occurs when two or more organisms place conflicting requirements on a limited supply of food, light, water or other resources.

consumer An organism which feeds upon other organisms (plant or animal).

continental drift The concept that the continents float on a molten mantle; that they have fragmented from earlier larger land-masses and have changed their relative positions.

cuticle Superficial layer of cells covering an animal or plant, the main purpose of which are protection and/or the reduction of water loss.

cytoplasm All the substance of a cell excluding its nucleus.

deciduous Plants that shed all their leaves in response to certain seasonal environmental changes, e.g. annual cold or drought.

decomposer An organism which breaks down dead organic matter, absorbing nutrients for its own growth and also releasing nutrients to the environment for use by other organisms. Many bacteria and fungi are decomposers.

density-dependent factors Factors which control the growth of a population and exert their effects more strongly as the populations become more dense; many diseases are examples of such a factor, and food supply might in some circumstances be another.

desertification The process by which an area previously marginal to a desert becomes true desert: climatic factors may be involved, as may human use of the area.

diversivore An animal whose diet includes both plant and animal material.

dominants The most characteristic species in a particular plant community and which largely controls the energy flow and the environment of all other species in a community. It is often the largest or strongest plant in the community, e.g. the oak tree in an oak wood.

dormancy A resting condition, when an organism is alive but slows or reduces the rate of the life processes.

ecology The scientific study of the relationships between organisms and both their living and non-living environments.

ecosystem A community of organisms interacting with one another and with the environment in which they live. The basic study unit of ecology, e.g. a pond, a salt marsh or a forest.

endemic A population of organisms restricted to a particular habitat or geographical range.

ephemerals Plants with a very short life-cycle which can complete this cycle during a temporary phase of favourable conditions, such as rainfall in a desert.

epiphytes A plant growing on another, not parasitically but using it merely for support, e.g. lichens and mosses on trees.

euphotic Well illuminated: applied to the surface zone of the sea (approximately upper 200 m) into which enough light penetrates for active photosynthesis.

eustatically emergent Applied to areas of land which become exposed as the general sea level falls, e.g. as water was taken in to form the ice sheets of glacial times.

exclosures Area where animals have been experimentally excluded from the vegetation to study their effect on plants, soils and other animals. The technique may also be applied to areas where humans exert a strong influence on an ecosystem.

feedback The return of some of the output of a system as an input, so as to exert some influence on the system.

food chain A sequence of organisms in which each is the food of a later member of the sequence, e.g. grass → sheep → man.

food web A number of feeding relationships which form a complex web rather than a simple chain.

gene A unit of genetic material localised in the chromosone and concerned with the development in the offspring of hereditary characteristics.

genetics That part of biology concerned with the study of heredity and variation.

genotype The genetic constitution of an organism; the total of all the genes present in an individual.

habitat The area of external environment in which a plant or animal lives.

halophytes Plants adapted to growth in salty soil, e.g. plants that grow well in river estuary conditions, or in deserts whose soils have a high salt content.

herbivore An animal that only eats plants.

hibernation When an animal passes the winter in a dormant state, i.e. the life processes are greatly slowed down.

invertebrates All animals without a backbone or spinal column.

leaching Removal of soluble salts from the soil, especially the surface layers, by water moving through the soil.

marsupial mammals Mammals found in Australia and the Americas and which shelter their young (born in an undeveloped state) in a pouch on the mother.

microflora Microscopic organisms, e.g. the fungi, small algae and bacteria of an area.

mineral nutrients Naturally occurring chemical elements or compounds used as food substances by plants, e.g. potassium, sodium, calcium.

morphology The study of form and how it develops, i.e. the external shape of an animal or plant and the course of its development.

mycorrhizal Fungi which grow on or around the root of a plant and form an association from which both benefit.

natural selection Processes occurring in nature which result in the survival of those individuals which are best fitted to their environment, and the passing on of their characteristics to the next generation.

net primary productivity The amount of organic matter formed by photosynthesis in excess of that used in respiration for a given area, i.e. the net addition of plant material per unit area per unit time.

nucleic acid Long-chain molecules which are the basic building blocks of genetic material.

organic matter Matter which has lived or is living.

pastoralism A way of life in which a human group subsists upon a domesticated animal, using it for food and probably for clothing, shelter and many material objects. Seasonal movement of the animal herds is frequently practised.

permafrost Any part of the Earth's surface layers continuously maintaining a temperature of 0°C or below for two or more years.

photosynthesis In green plants the production of carbohydrates from water and carbon dioxide using energy absorbed by chlorophyll from sunlight.

physical structure The way in which an ecosystem appears in section, e.g. the number of tree layers and shrub layers, the presence of grass and herb layers and the type of soil.

physiognomy The general appearance of an ecosystem or a community as characterised by its life forms.

physiology The study of the biochemical processes of living organisms.

phytogeography Plant geography is the study of the spatial distribution of plants.

phytoplankton The microscopic floating plant life in the sea.

phytotoxic These are chemical substances given off by plants which inhibit the growth of other plants.

predation The act of capturing and feeding by a member of one population on another, usually applied to the eating of plants by herbivores and the eating of herbivores by carnivores.

producer An organism able to manufacture food from simple inorganic substances, i.e. green plants and chemosynthetic micro-organisms.

protein Complex molecules characteristic of living matter and consisting of aggregates of amino acids.

pyrophytes Plants with an unusually high resistance to fire, e.g. in the form of very thick bark, or seeds which germinate only after having been heated by fire.

realms Regions of the Earth's surface having assemblages of plants and animals characteristic of that region, e.g. the marsupial fauna of the Australasian faunal realm.

respiration The breakdown of organic compounds to release energy in living cells.

sessile Of animals, fixed to one spot, sedentary, not able to move about.

secondary production In an ecosystem, the quantity of animal tissue which grows per unit area per unit time. It may come from either herbivores or carnivores, or from animals of the decomposer layer.

species The smallest unit of biological classification commonly used. The individuals have major characteristics in common; for most animals and many plants a species is roughly a group of individuals able to breed amongst themselves but not with organisms of other species groups.

species diversity The number of different species found in a community or region.

subspecies A group whose members resemble each other in certain characteristics and differ from other members of the species, although there may be no clear dividing line. The differences are often due to partial or spatial isolation.

succession Progressive change in composition of an ecosystem towards a largely stable state, even though there is no change in external environmental conditions.

taxon General name for the categories into which organisms are classified, e.g. species, genus, family, order, class, phylum, kingdom.

tolerance The range of conditions under which a plant or animal can survive and reproduce.

trophic level The designation of groups of organisms in an ecosystem according to their food sources. The first trophic level consists of plants, a second trophic level of herbivores, and so on. A terrestrial ecosystem will normally contain three or four such levels and a decomposer level.

unstable soils Soils which are prone to constant or intermittent movement downslope. They are common on areas of high slope and also during periods of alternate freezing and thawing.

variation The range of characteristics within a taxonomic group, the magnitude of which is insufficient to cause it to be classified differently, e.g. four-leaved clovers are still clovers, Manx cats are still cats.

zoogeography The study of the spatial distribution of animals.

Index

Italic page numbers refer to text figures and bold page numbers refer to tables.

agriculture 42, 47, 51-2, 55, 64, 66-9, 72, 74-7, **76**, 84, 86-7
  origins of 69-70, 81
aquatic environments 5-7, 55-64
attitudes to nature 91

biogeographical cycles (*see also* mineral nutrients) 21-3, *21-2*, 26-30, 55
biological control 23, 76-7, 81
biological productivity 17-19, 30-1, 37-8, 40-1, 42-4, 46-7, 48-9, *49*, 54-61, 66
  secondary productivity 31, 54, 58
  of agriculture *20*, *30*, 74, **76**
biomass 18, 22, 29-30, 41, 44-6, 49-52, 54-5, 60
biomes 32-64, *33*, *62*

calcicole 5, *5*
calcium carbonate 5
carbohydrate 18
carbon 17
carbon dioxide 6, 18
carrying capacity 25-6, 91
chaparral – *see* sclerophyll ecosystems
chemical control 41-2, 76, 87-8
chlorophyll 4, 18, 21
community vi, 7, 14-16
competition 7
conservation *60*, 80, 83-9, 91
contamination 77-8, *77-8*
continental drift 11
coppice 47, *51*, 52
coral reefs and islands 30, 55-6, 60-1, *60*

decomposition 21-2, 30, 51, 64
desertification 36
deserts *9*, 32-6, *34-5*
'development' 91
distribution (of species) *9-10*, 10-16
  continuous 10
  discontinuous 10-16, *11-12*
  endemic 11, 80
domestication and domesticated species 36, 55-6, 57-8, 66-74, 66, *68-71*, 84
dominants *15*, 15-16, 46-9, 53
dormancy 49
dunes 35-6

ecology vi, 16
ecosystems
  in general vi, 65-6, *68*, 72-3
  as concept vi, 16, *17*
  diversification of 80-3
  management of 86-9, *86-7*
  mature or climax 26-30
  mosaics *29*, 30
  obliteration of 78-80
  processes 17-31
  protection of 85-9
  simplification of 66, 74-8, *66*

energy vi, 5, 16, **66**, 74-5, **76**, 90
  calorific value of 19-20
  chemical 5, 18
  density 75, **76**
  flow in ecosystems 17-20, *19-21*, 25, 51, 60
  from fossil fuels 65, 67, 71, 74
  from wind and water 66
  nuclear 65, 67
ephemeral plants 6, 34-6
estuaries 60-1, 64
evolution 3-5, 12-13, 44-5, 56, 65
extinction 55, *60*, 78-80, *80*

faunal regions 13-15, *15*
fire 7, 9, 39-40, *45*, 45-9, 52, 55, **62**, 65-6, 74-5, 88
  fire tolerant plants (pyrophytes) 7, 47
food chain 19, *19*, 27, 29, 58, 74
food production 31
food web 19, *20*, *57*, *61*
forests
  boreal coniferous *47*, 47-9
  destruction of 52, 55, 61, 72-3, 77
  'primary' 52
  temperate deciduous 49-52, *50*
  tropical evergreen 53-5, *53*
floral regions 11, 13

garrigue – *see* sclerophyll ecosystems
genetics vi, 9, 30, 65-9, *66*, *71*, 75-6
  modern manipulation of 71-2
germination 9, 36
grasslands 40-2, *40*, *42*
grazing 9, 36, 39, 41-2, 43-4, 47, 52, 55-6, 80-1, 86
  overgrazing 73

habitat vi, 15-16
halophytic plants 7, 61
heat 18-19, 36
hibernation 5, 5
hunting 39-40, 49, 52, 59, 66
hydrogen 20-1
Hymns ancient and modern (revised) 32

industrialization 67, 74-5
introductions 56, 80-3, *81*, *83*, 88
islands 11, 39, 55-6, *60*, 80-1

leaching 7, 22-3, 54
light 5, 9, 47, 50, 53, 56
litter 22-3, *23*, 41, 48-9, 51, 53-5

mallee scrub – *see* sclerophyll ecosystems
maquis – *see* sclerophyll ecosystems
man
  and biogeographical processes 65-91
  animal characteristics of 65
  his attitudes to nature 91
  populations of 26, 91

migration 38, 41-2, 44, 51
mineral nutrients (*see also* biogeochemical cycles) 7, 9, 20-3, 41-2, 49, 51-2, 54-8, 61, **62**, 67, 74
morphology 3
mycorrhizae 54

natural selection 1, 7, 9-10, 67-8, 84
net primary production (NPP) – *see* biological productivity
nitrogen 20-3, 42, 54, 58, 77

oceans 5, 31, 55-61, **56**, *57*, *61*, 64
oil – *see* petroleum
oxygen 6, *7*, 59, 61, 77

parasites 7, 22, 82
pastoralism 36, 42, 47, 64, 73
permafrost 37
petroleum (*see also* fossil fuels) 7, 38-9, 75
pets 82-3
physical environment 4-7, 15-16, 55, **62**
photosynthesis *5*, 5-6, 18, 56, 65
phytogeography vi, *9-13*
pollarding *51*, 52
populations 10, 23-6
    competition within 23
    density-dependent 25
    dynamics of human 26, *26*, 91
    oscillations of *15*, 38, 60
    parameters of 23, *24*
    regulations of 24-6, 91
predation 7, 22, 25

realms (zoogeographical regions) 13
recreation (and tourism) 36, 84, 88-9, *89*
respiration 18

salt marshes – *see* estuaries
savanna, tropical 42-6, *44-5*
scientific research 83-4
sclerophyll ecosystems 46-7
seas – *see* oceans
soils vi, 5-7, 10, 21-3, 32, 37, 39, 41-2, 45-6, 49-52, 54-5
    erosion 76-7
solar radiation 4-5, 23, 30, 54
species
    as concept 1, 4, 7
    as resources 91
    number and diversity of *2*, *5*, *6*, 12-15, 43, 46, 50, 53-6, 78, 80, 84
    protection of 85-6
stability 15, 19, 84
succession *26*, 27, 29-30, 48, 54, 74
sugar 18
sulphur 7, 21
    dioxide 77

temperature *3*, 5, 32, 36-7, 42, 47, 53-5, **62**, 77
timber production 52
tolerance 4, *4*, 7
tourism – *see* recreation
transpiration 34, 47
trophic levels 19-20, *19-21*, 57
tundra 36-9, *38-9*

variation 1, *1*, 4, 77

water 5-7, **6**, 30, 32-7, 40, 41-2, 43, 47, 49, 53, 54, **62**, 86
weeds 24

zoogeography vi, *14-15*